Mechanism and Design of Sequencing Batch Reactors for Nutrient Removal

Scientific and Technical Report Series

Mechanism and Design of Sequencing Batch Reactors for Nutrient Removal

Scientific and Technical Report No. 19

Nazik Artan and Derin Orhon

Istanbul Technical University, Environmental Engineering Department

Published with the knowledge of the IWA Specialist Group on Nutrient Removal and Recovery

IWA Publishing

Published by IWA Publishing, Alliance House, 12 Caxton Street, London SW1H 0QS, UK
Telephone: +44 (0) 20 7654 5500; Fax: +44 (0) 20 7654 5555; Email: publications@iwap.co.uk
Web: www.iwapublishing.com

First published 2005

Index prepared by Indexing Specialists, Hove, UK.
Printed by Lightning Source

Disclaimer

The information provided and the opinions given in this publication are not necessarily those of IWA or of the IWA Specialist Group on Nutrient Removal and Recovery, and should not be acted upon without independent consideration and professional advice. IWA and the authors will not accept responsibility for any loss or damage suffered by any person acting or refraining from acting upon any material contained in this publication.

British Library Cataloguing in Publication Data
A CIP catalogue record for this book is available from the British Library

Library of Congress Cataloging- in-Publication Data
A catalog record for this book is available from the Library of Congress

ISBN 1843390825

Contents

Preface

This report is prepared to provide the basic understanding and description of a unified basis of design for suspended-growth *sequencing batch reactors*. It is hoped that it will be useful as a complement to an earlier volume of the IWA's Scientific and Technology Report Series, mainly covering practical aspects of *sequencing batch reactor technology*, but also emphasizing the need for appropriate design guidelines.

The proposed basis for design is elaborated in a stepwise approach for both organic carbon removal and nutrient control. Nutrient removal from wastewaters presents a vital importance for the protection of sensitive water bodies from eutrophication. Greater concern for the water quality has been the major incentive for the implementation of stringent effluent regulations. The increasigly stricter nitrogen and phosphorus limits on wastewater discharges have stimulated studies on improving the activated sludge process and developing single sludge biological nutrient removal process configurations. Accumulated experience indicate that SBR also incorporates all the necessary attributes to qualify as a viable and competitive alternative to continuous-flow systems for nutrient removal. With their intrinsic flexibility of operation, SBR systems can be engineered in a simple way, to provide the most feasible process options for different nitrogen and phosphorus removal schemes. The striking feature of various SBR configurations for nutrient removal is that they can all be interpreted with a unified design approach, provided that the specific operating functions are fully understood and interpreted in terms of related mass balance equations.

The main message of the report is that SBR systems with different modes of filling and operation should not be viewed and interpreted as different generic SBR groups. The unified design procedure, when properly understood and applied, will provide all the necessary tools for the selection of the most appropriate cyclic operation scheme, the sequence of necessary phases and filling patterns for the particular application. Support of model simulation and evaluation as an integral component of system design is also emphasized.

The report has greatly benefited from the accumulated scientific work so far conducted at Istanbul Technical University on microbial kinetics and especially, on the mechanistic evaluation of different activated sludge configurations.

We gratefully acknowledge the contribution of Reha Artan on practical aspects of SBR operation. We would also like to express our gratitude to our colleagues, Güçlü Insel and Nevin Özgür Yağcı for their valuable time and effort in the preparation of various parts of the manuscript.

Nazik Artan
Derin Orhon

The Authors

NAZIK ARTAN

Nazik Artan is Professor of Environmental Sciences and Engineering at Istanbul Technical University. She received a B.S. degree in chemical engineering, and M.S. and Ph.D. degrees in environmental engineering from the same university. Currently, she acts as coordinator of the graduate program on Environmental Biotechnology.

Her studies concentrate on process modeling, SBR technology and biological nutrient removal. She has published more than 50 papers in scientific journals and is a member of the Turkish National Committee on Water Pollution Research and Control.

DERIN ORHON

Derin Orhon is Professor of Environmental Sciences and Engineering and Dean of the School of Civil Engineering at Istanbul Technical University. He received an M.E. degree in civil engineering from the same university, and M.S. and Ph.D. degrees in sanitary engineering from the University of California at Berkeley.

Prof. Orhon received the Science award of the Scientific and Research Council of Turkey in 1998. In 2000, The International Water Association (IWA) awarded him with the Samuel H. Jenkins Medal. In the same year, he was elected full member to the Turkish Academy of Sciences and in 2001, to the Science Board of the Scientific and Research Council of Turkey.

His principal reseach interests focus on environmental biotechnology, process kinetics and modeling, biological nutrient removal and industrial waste management with specific emphasis on agro-industries.

Prof. Orhon has so far authored more than 150 papers in scientific journals. He also serves as consultant to both national and international institutions and private concerns. He acts as chairman of the Turkish National Committee on Water Pollution Research and Control.

Nomenclature

η	Anoxic correction factor	-
μ_{Amax}	Maximum autotrophic growth rate	[1/day]
μ_{Hmax}	Maximum heterotrophic growth rate	[1/day]
β	The last fill fraction	-
ΔP	P removal capability	[mgP/l]
θ_h	Hydraulic retention time	[h]
θ_X	Total sludge age	[day]
θ_{XA}	Aerobic sludge age	[day]
θ_{XE}	Effective sludge age	[day]
θ_{XP}	Process sludge age	[day]
AOR	Actual oxygen requirement per hour	[mgO_2/l.h]
b_A	Autotrophic endogenous decay coefficient	[1/day]
b_H	Endogenous respiration rate coefficient	[1/day]
b'_H	The decay coefficient in death-regeneration model	[1/day]
C_{i1}	Influent substrate concentration	[mgCOD/l]
$C_{N,E}$	Allowable total nitrogen concentration	[mgN/l]
C_{S1}	Influent total biodegradable COD	[mgCOD/l]
C_{T1}	Influent total COD	[mgCOD/l]
C_{TKN1}	Influent Total Kjeldahl Nitrogen	[mgN/l]
E_P	Potential denitrification efficiency	[%]
f_{CS}	Biodegradable fraction of COD	-
f_E	Inert fraction of biomass	-
f_P	Particulate inert fraction of biomass	-
f_{ES}	Soluble inert fraction of endogenous decay	-
f_{EX}	Particulate inert fraction of endogenous decay	-

f_{FS}	Fixed solids/total COD of influent	[gTSS/gCOD]
f_{SS}	Readily biodegradable fraction	-
FTR	Fill time ratio	-
f_{XI}	Particulate inert COD fraction	-
i_{NBM}	Nitrogen content of biomass	[gN/gCOD]
i_{NSI}	N content of S_I	[gN/gCOD]
i_{NXI}	Nitrogen content of X_I	[gN/gCOD]
i_{PBM}	P fraction of biomass	[gP/gCOD]
i_{PXI}	P content of X_I	[gP/gCOD]
i_{TSSCOD}	Coefficient to convert COD to TSS	[gTSS/gCOD]
K_{h1}	Maximum hydrolysis rate for S_H	[1/day]
K_{h2}	Maximum hydrolysis rate for X_S	[1/day]
K_La	Oxygen transfer coefficient	[1/day]
K_{NH}	Half saturation coefficient	[mgN/l]
K_S	Half saturation coefficient	[mgCOD/l]
K_{X1}	Half saturation coefficient for S_H	[gCOD/gCOD]
K_{X2}	Half saturation coefficient for X_S	[gCOD/gCOD]
m	Number of cycles per day	-
MLSS	Mixed liquor suspended solids	[mg/l]
n	Number of reactors	-
N_A	Available nitrate	[mgN/l]
N_{DP}	Denitrification potential	[mgN/l]
N_{ER}	Nitrate nitrogen equivalent of the electron acceptor demand for endogenous respiration	[mgN/l]
N_{OX}	Nitrification capacity	[mgN/l]
N_{SND}	N denitrified in aerated period	[mgN/l]
N_{SS}	Nitrate nitrogen equivalent of the electron acceptor demand for heterotrophic growth on S_S	[mgN/l]
N_X	Nitrogen in biomass	[mgN/l]
N_{XS}	Nitrate nitrogen equivalent of the electron acceptor demand for heterotrophic growth on X_S	[mgN/l]
OR	Daily amount of oxygen requirement	[kgO_2/day]
OR_H	Heterotrophic oxygen requirement	[kgO_2/day]
OR_T	Total oxygen requirement	[kgO_2/day]
P_{cont}	Phosphate content of the sludge	[gP/gTSS]
P_{XH}	Heterotrophic biomass produced per unit time	[g/day]
P_{XT}	Sludge production	[gTSS/day]
Q	Daily volumetric flow rate	[m^3/day]
R	Recycle ratio	-
R_N	Nitrate recycle ratio	-
r_{OT}	Oxygen transfer rate	[$gO_2/m^3.h$]
r_S	Observed conversion rate for C_S	-
R_S	Sludge recycle ratio	[%]
S_A	Readily biodegradable COD, as VFA	[mgCOD/l]
S_e	Effluent soluble COD	[mgCOD/l]
S_F	Safety factor	-
S_H	Soluble slowly biodegradable COD	[mgCOD/l]
S_{HI}	Influent readily hydrolysable COD	[mgCOD/l]
S_I	Inert soluble COD	[mgCOD/l]

S_{I1}	Influent inert soluble COD	[mgCOD/l]
S_{NH},	Effluent ammonium nitrogen	[mgN/l]
S_{NO}	Effluent nitrate nitrogen	[mgN/l]
$S_{NO,D}$	Design effluent nitrate nitrogen concentration	[mgN/l]
S_O	Dissolved oxygen concentration	[gO_2/m^3]
S_{OS}	Saturation conc. of dissolved oxygen	[gO_2/m^3]
S_{PO4}	Soluble phosphorus concentration	[mgP/l]
S_S	Readily biodegradable COD	[mgCOD/l]
S_{S1}	Influent readily biodegradable COD	[mgCOD/l]
S_T	Total soluble COD	[mgCOD/l]
S_{T1}	Influent total soluble COD	[mgCOD/l]
SVI	Sludge volume index	[ml/g]
T_A	Duration of aeration period	[h]
T_{AN}	Duration of anaerobic period	[h]
T_{AOX}	Duration of anoxic period	[h]
T_C	Total cycle time	[h]
T_D	Duration of draw phase	[h]
T_E	Effective time	[h]
T_F	Duration of fill phase	[h]
T_I	Duration of idle phase	[h]
T_M	Duration of mixing period	[h]
T_P	Process time	[h]
T_R	Duration of react phase	[h]
T_S	Duration of settling phase	[h]
T_{S+D+I}	Duration of settle, draw and idle phases	[h]
V	Reactor volume at a given time, t	[m^3]
V_0	Initial volume	[m^3]
V_F	Fill volume	[m^3]
V_S	Settled sludge volume	[m^3]
V_T	Total reactor volume	[m^3]
X_E	Particulate inert COD	[mgCOD/l]
VER	Volumetric Exchange Ratio	-
X_{FS1}	Influent fixed solids concentration	[mgCOD/l]
X_H	Active heterotrophic biomass	[mg/l]
X_I	Inert particulate COD	[mgCOD/l]
X_{I1}	Influent inert particulate COD	[mgCOD/l]
X_{PAO}	Phosphate accumulating organisms	[mgCOD/l]
X_{PHA}	Stored poly-hydroxy-alkanoates	[mgCOD/l]
X_{PP}	Stored poly-phosphate	[mgCOD/l]
X_R	Achievable concentration in the recycle flow	[mg/l]
X_S	Slowly biodegradable COD	[mgCOD/l]
X_{S1}	Influent slowly biodegradable COD	[mgCOD/l]
X_T	MLSS concentration in the reactor	[mg/l]
X_{T1}	Influent total particulate COD	[mgCOD/l]
X_{TSS}	Total suspended solids	[mgTSS/l]
Y_A	Autotrophic yield coefficient	[gCOD/gN]
Y_H	Heterotrophic yield coefficient	[gCOD/g COD]
Y_{NA}	Net autotrophic yield	[gCOD/gN]
Y_{NH}	Net heterotrophic yield	[gCOD/gCOD]

1

Introduction

1.1.HISTORICAL PERSPECTIVE

Activated sludge is the most ingeniously conceived process today for the biological treatment of domestic and industrial wastewaters. For almost a century, it has been the traditional scheme for the effective removal of organic carbon. In the last few decades it has been successfully explored and tested for its potential of nitrogen and phosphorus removal. It has also served as the platform for the most remarkable and productive research effort in environmental sciences, constantly improving and updating the understanding of the fundamental microbial mechanisms involved, which readily translated into process improvement.

Today, continuous-flow activated sludge systems almost always guarantee satisfactory performance with minimal supervision and maintenance for conventional wastewaters. Despite substantial practice and accumulated experience however, more and more stringent effluent quality requirements often imposing removal of nutrients together with organic carbon and the increasing complexity of wastewaters seriously challenged their efficiency. The difficulties encountered in the operation of the conventional process configuration have been the basic motivation of extensive efforts to review existing design criteria and operation practices as well as to develop process modifications likely to offer improved treatment potential. While the problems were recognized, they were not always thoroughly and scientifically defined, so that rational corrective measures could not always be taken. Therefore, the majority of plant improvements and modifications had to rely on experience and engineering judgement. New ideas were mostly developed by trial and error for their

 Mechanism and Design of Sequencing Batch Reactors for Nutrient Removal: Scientific and Technical Report No. 19 by Nazik Artan and Derin Orhon. Published by IWA Publishing, London, UK. ISBN: 1843390825.

particular situation. Several new concepts and process modifications emerged from these individual efforts and found acceptance in wastewater treatment practice.

The *sequencing batch reactor* (SBR) is perhaps the most promising and viable of the proposed activated sludge modifications today for the removal of organic carbon and nutrients. In a relatively short period, it has become increasingly popular for the treatment of domestic and industrial wastewaters, as an effective biological treatment system due to its simplicity and flexibility of operation.

The SBR is basically a single tank that serves both as a biological reactor and settler in a temporal sequence; whereas aeration and settling are simultaneous but in a spatial sequence in the continuous-flow activated sludge, they are carried out in the same reactor but in a temporal sequence in the SBR systems. SBR inherently involves a cyclic operation, each cycle incorporating the same pattern of successive selected phases. As contrasted to continuous-flow systems, wastewater feeding is provided during the desired portions of the cycle, which then secures a *batch wise, fill-and- draw* type of an operation.

The fill-and-draw scheme, constituting the basic principle of the SBR system was the essential instrument that initiated the development of the original activated sludge, an excellent review of the historical evolution of the activated sludge process was recently presented by Wilderer *et al.* (2001). The experiments of Arden and Lockett (1914) should be recognized as the major pioneering work leading to this most widely used process in wastewater treatment. Their work was essentially a follow-up of the earlier experiments in Massachusetts, which demonstrated that aeration of sewage for a short time in a tank containing slabs of slate about 2-3 cm apart produced a "compact brown growth" on the slate and a clear effluent. In their fill-and-draw experiments, they saved the flocculent solids which they called *activated sludge* and reported that their repeated use in sewage treatment greatly improved the purification potential of simple aeration; the improvement was observed to be closely related upon the ratio of activated sludge to the sewage treated. They also concluded that a clear, *well-oxidized* effluent could be obtained after aeration of sewage in contact with activated sludge, for a period of six to nine hours – a period adopted as a design parameter in actual practice. Similar fill-and-draw experiments conducted the same year at the University of Illinois (Mohlmann, 1917) confirmed the function of the activated sludge as an indispensable factor for sewage stabilization and a six-hour aeration period for a satisfactory treatment. It was also reported that the aeration rate should be between 7 to 11m^3 of air/m^3 of sewage – another operational value that has survived the years as a key design parameter.

The discovery of *activated sludge* found immediate recognition in practice, both in England and in the US. Shortly after the publication of the results obtained by Arden and Lockhett, a treatment plant operated as a fill-and-draw system was installed for Daryhulme, Manchester (Orhon and Artan, 1994). The first similar fill-and-draw plant in the US was put into operation in 1915, in Milwaukee, Wisconsin. From 1914 to 1920 several full-scale fill-and-draw systems of different size were placed into operation in England and in the US Wilderer *et al.* 2001).

The performance of the pioneer activated sludge plants in their early years of operation was quite satisfactory. They provided a number of basic criteria for the principal design features of new plants and firmly established the activated sludge process. After 1920 however, taking advantage of the flocculent properties of activated sludge along with its "clarifying power", efforts were then directed towards the adaptation of the process to operate under continuous-flow conditions. Successful experience with this new approach, together with the promotion of the diffused air process as a feasible means of air provision consolidated the continuous-flow principle as the major practical method for activated sludge operation.

After the choice for continuous-flow processes, the interest for batch-fed systems was revived in the 1960s with the development of new technology and equipment. The system in its newly devised scheme with a periodic influent feeding and periodic discharge was called the *sequencing batch reactor*. The following research mainly focused on exploring and stressing the advantages of the SBR over the conventional continuous-flow system. Among major examples of this research effort is the study by Dennis and Irvine (1979) where sludge settleability characteristics were found to vary markedly with different fill/react ratios. Hopker and Schroeder (1979) showed that the lower feed strength resulted in better effluent quality, and semi-batch treatment was more suitable in minimizing dispersed growth. Irvine et *al.* (1979) demonstrated the feasibility of nitrification-denitrification, given proper design and operation. Ketchum and Liao (1979) explored the potential of the SBR for phosphorus removal. The results of these studies secured the credibility of the SBR as a reliable process contributed to its acceptance in practice as a viable and resourceful alternative for the biological treatment of domestic and industrial wastewaters.

1.2. CURRENT EXPERIENCE

The most striking feature of the SBR process is the system flexibility offered with a very simple physical structure. The operation alternatives inherently associated with SBR has been the basis of its promotion both in full-scale application and as the perfect tool for fundamental and applied research.

1.2.1. Basic and Applied Research

The SBR process has the mechanistic capability to serve as a perfect experimental instrument for the exploration of the intricate array of microbial mechanisms associated with different processes involved in biological wastewater treatment. Conceptually, it offers a much better opportunity to follow the functional relationships between the relevant parameters of a selected process as compared to a continuous-flow, completely mixed reactor, (CSTR). In the latter, each parameter can only be observed as a single value at a given steady-state operation; this value reflects the combined resulting effect of all the biochemical mechanisms involved. The relative importance or even the presence of any of these mechanisms cannot possibly be identified. However, the SBR reflects during each cycle the transient responses of all the observed parameters, such as COD, nitrogen forms, biochemical storage products, oxygen uptake rate, etc. Moreover, the SBR is also much better suited for this purpose compared with a simple batch reactor, mainly because it is operated at steady-state with identical responses in each cycle, reflecting this way the microbial culture history for selected operating conditions, a very important factor often overlooked in batch experiments.

Appropriate interpretation and modelling of a number of fundamental issues dealing with different aspects of biological treatment had to rely on the SBR process for experimental support. The study of Orhon *et al.* (1986: 1989) may be remembered as one of the first examples highlighting the merit of the SBR as a valuable research tool, where a model proposed for the formation of soluble residual microbial products, an important parameter especially for the treatment of industrial wastewaters, was calibrated and verified for a set of experimental data, accurately predicting COD accumulation at the end of each cycle. Among recent research efforts is the work by Karahan *et al.* (2003), which studied the biodegradation of different types of starch as typical hydrolysable substrates, through observation and modelling of glycogen concentration and oxygen uptake rate profiles during a cycle at steady-state; they suggested a new model structure involving both simultaneous growth and biochemical storage. Similarly, based on extensive experimental data collected from an SBR system fed with acetate as the sole carbon source at different COD/P ratios, Yagci *et al.* (2003) proposed a new metabolic model for acetate uptake by a mixed culture of phosphate and glycogen accumulated microorganisms under anaerobic conditions.

The SBR has also been extensively utilized, both at laboratory and pilot-scale for the experimental assessment of the level of removal efficiencies achieved in the treatment of domestic and industrial wastewaters, an elaborate review of various applications of the SBR technology was presented by Mace and Mata-Alvarez (2002). In these studies, system flexibility of the cyclic operation offering additional system parameters such as cycle time, fill time, etc., mostly selected at random, have been elaborated as substantial competitive advantages of the SBR against the continuous-flow configuration. Most of the studies were conducted with a pre-selected set of operating parameters and almost always reported satisfactory removal performance for the particular case of application.

Application of the SBR technology for improved treatment of domestic sewage has been a major focus of attention. The wide spectrum of the operating parameters selected for this purpose should be underlined. Imura *et al.* (1993) chose to adopt a 4 cycles/day (6-h cycles) operation for complete nitrification and BOD, SS and total phosphorus removals of over 95%. Umble and Ketchum (1997) obtained similar results with a 12-h cycle operation for effluent nutrient recovery through agriculture. Bernardes and Klapwijk (1996) reported satisfactory nutrient removal with a set of two SBRs in series, each operated at 6 cycles/day and a different sequence of arbitrarily selected phases within the cycle. The SBR was also tested as quite a successful biological treatment option for landfill leachates, again with a wide array of recommended operating parameters involving a range of 1.5 – 10 days for the hydraulic retention time, 20 – 50 days for the solids retention time and 12 – 24 h for cycle time (Dollerer and Wilderer, 1996; Zaloum and Abbot, 1997; Timur and Ozturk, 1997; Yilmaz and Ozturk, 2001).

The SBR is also widely explored and tested for the treatment of industrial wastes. Satisfactory and reliable performance is reported for winery wastewaters (Torrijos and Moletta, 1997), brewery wastewaters (Ling and Lo, 1999), food industry wastewaters (Raper and Green, 2001) dairy wastewaters (Mohseni and Bazari, 2000) slaughterhouse wastewaters (Belanger *et al.*, 1986) piggery wastewaters (Lee *et al.,* 1997) pulp and paper mill effluents (Franta and Wilderer, 1997) and tannery effluents (Carucci *et al.,* 1999). Similar to the experience with domestic sewage, very little evidence, if any, is available in related literature to justify the specific operating parameters and design criteria selected to ensure the desired performance of the SBR system for the particular industrial wastewater. A remarkable exception to this statement is the comprehensive evaluation of the SBR treatment of tannery wastewater by Murat *et al.* (2002) involving appropriate wastewater characterization and COD fractionation of the wastewater for modelling, experimental assessment of model coefficients, mass balances for carbon and nitrogen removal, performance prediction of SBR operation based on process stoichiometry and model calibration of system performance. Effects of temperature were investigated in a following study (Murat *et al.,* 2004), which established the nitrogen balance of the SBR system treating tannery wastewater for a wide temperature range between 9 and 30° C and evaluated the experimental results by means of model calibration of COD, nitrate and ammonia nitrogen concentration profiles during cyclic operation.

1.2.2. Full-scale Application

The urge for simpler and more reliable treatment systems, the comfort of system flexibility and additional operating parameters and extensive applied research highlighting SBR as a novel process with competitive advantages against the conventional continuous-flow activated sludge have been the major ingredients for the increasing success of this process in full-scale application, both for domestic and industrial wastewaters.

The development of the Intermittently Decanted Extended Aeration, (IDEA), a system with continuous influent feeding but intermittent acration and decanting, between 1965 and 1975 in Australia (Goronszy, 1979), together with the conversion of the full-scale continuous-flow activated sludge treatment plant at Culver, Indiana, USA in 1980 (Irvine *et al.* 1983) may be cited as pioneering efforts that have contributed a lot to the success of the SBR process. The ability of the system to easily adapt to nutrient removal by simple incorporation of an anoxic/anaerobic phase has been an equally important factor contributing to the increasing popularity of the SBR systems. Today, full-scale application of the SBR process is wide spread with more than 1200 full-scale plants in North America; more than 700 plants in Japan, mainly as small installations in rural agricultural areas; around 150 plants in Germany with a large portion for industrial applications; more than 100 plants in Australia (Wilderer *et al.* 2001). In Turkey, close to 1000 sewage treatment applications exist, mostly package and standardized plants for effluents from holiday resorts and small communities, in addition to some 50 plants for a variety of industrial wastewaters. More than 20 of these industrial applications were reported (Artan *et al.,* 1996).

The application of the SBR process has not been limited with suspended growth systems. SBR alternatives using biofilm systems have also been developed with the generic name of *sequencing batch biofilm reactor*, (SBBR) and found application in the treatment of domestic and industrial wastewaters (Wilderer, 1992; Pujol *et al.,* 1998).

1.3. NEED FOR MODELING AND A UNIFIED BASIS FOR DESIGN

The development of the SBR process took place in a period that could also be associated with remarkable achievements in the interpretation of fundamental microbial mechanisms, which greatly affected the modelling, and the design of the activated process. Introduction of a number of substrate components with different rates of biodegradation, the concept of residual microbial products, the ensuing detailed COD fractionation in terms of different substrate and biomass parameters, incorporation of hydrolysis as a major process for slowly biodegradable substrate, identification of microbial storage products and their utilization under a sequence of feast and famine conditions may be cited among revolutionary landmarks drastically changing the conventional understanding of the system.

The promotion of the SBR however did not benefit from these conceptual achievements as it basically developed as a polishing effort of the original, sixty years old pioneering engineering idea, mainly for two reasons: Firstly, related applied research chose to concentrate mainly upon the competitive advantages offered by the operation of the SBR system against the conventional continuous-flow scheme, such as the elimination of the often problematic settling tank and equipment, ability for better operation control, etc.; it emphasized new features of the system such as intermittent influent feeding, cyclic operation, etc., and manipulation of these parameters like fill time, cycle time, fill/react ratio, etc., in different combinations, mainly as a supporting evidence of the advantages of the system. In short, the common attribute of most SBR studies was perhaps the effort to acquire the required engineering experience.

Secondly, there was a distinct effort to promote the SBR process as a novel, original biological treatment process rather than an alternative, although ingenious, activated sludge configuration. Even SBR systems with different feeding patterns have been treated as almost different processes. In fact, Wilderer *et al.* (2001) identified four generic SBR groups based on operation with different phases within a cycle (i) systems with a periodic influent, a react phase and an idle phase; (ii) systems with a periodic influent, a react phase and no idle phase; (iii) systems with interrupted influent, a selector and no react and idle phase and (iv) systems with a continuous influent. They may be considered as useful engineering options often resulting from physical and operational constraints but they should definitely not be conceived as generic groups offering selective treatment alternatives in the removal of organic carbon and removal. On the opposite, the critical issue perhaps obstructing the rational interpretation of the process has been the systematic differentiation of the fill and react phases. From a mechanistic viewpoint, they merely represent different operating conditions for the same biological reactor.

Similar to the continuous-flow activated sludge appropriate design of the SBR process has been a major issue of concern. Wilderer *et al.* (2001) acknowledge the lack of necessary design and operating guidelines for sizing the SBR and programming the control mechanisms often resulting in the failure of the system in meeting the required effluent standards but they chose not to provide a unified design approach because of the variety of reactor types. Consequently, the available design procedure is limited with the simplistic traditional approach of assigning selected volumetric and organic loadings and sludge ages values: In this context, EPA (1999) suggested a cycle time of 4 hours, a food to biomass (F/M) ratio in the range of 0.15 – 4.0 kg BOD_5/kg SS.day and a hydraulic retention time of 6 to 14 hours as key design parameters for conventional domestic wastewater treatment; much wider ranges of no practical value were recommended for industrial wastewaters, similar values were also indicated by Kirschenman and Hameed (2000). This type of a design method, although useful to a limited extend, is clearly inadequate and may at times be misleading in the sense that with such design parameters adopted, an independent arbitrary selection of other parameters inherently associated with SBR can be exercised.

1.3.1. The concept of process stoichiometry and modelling

The SBR process offers a great flexibility and variety of operation. The abundance of possible operation alternatives with SBR makes it an important task to develop a rational and scientifically sound design approach. This flexibility of operation, if well understood and interpreted in terms of governing biochemical processes, may prove very useful. The past experience however has developed mainly as an engineering exercise of trial and error where different operation options have been experimentally tested without so much emphasis on process kinetics and stoichiometry.

The performance of the SBR is now fully interpreted in terms of basic principles incorporated into recent activated sludge models. For such evaluations, the process offers the advantage of observing and modelling concentration transients for selected key parameters such as COD, N forms etc. It also provides the necessary flexibility of operation to transmit the outputs of kinetic evaluations into application, by appropriate adjustment of cycles and manipulation of aerated and non-aerated phases.

Despite accumulating information, a unified basis for design is still lacking, a situation regarded by many as the major obstacle hindering broader practical application of the SBR. Principles of a systematic approach to the design of SBR can be formulated similar to those for continuous-flow systems. Though the possible design and operation choices seem to be abundant, steady state SBR design applies the same mass balances. However, SBR has inherent constraints besides some advantages and requires different design considerations. While a simple approach may be tolerated for the organic carbon, nitrogen and phosphorus removal require a unified basis for design and the support of process modelling.

In recent years, principles of a systematic approach for the SBR design have been defined for using basic stoichiometry in a way that allows comparative evaluation with continuous-flow activated sludge. Different steps in the SBR design may be interpreted in quite a similar approach provided that its specific operating features are fully understood and interpreted in terms of related mass balance equations. Orhon and Artan (1994) first used basic COD mass balance relationships for the mechanistic design of SBR systems. Artan *et al.* (2001) proposed a rational dimensioning approach for nutrient removal by SBR, based on scientific information on process stoichiometry, modelling and practical constraints in design and operation. Tasli *et al.* (2001) used a similar approach in retrofitting existing SBR systems for nutrient removal, where the efficiency of retrofitting was found to rely on the balance between the sludge age, the initial settled sludge volume in the reactor and the ratio of the initial volume to the feed volume in each cycle. Artan *et al.* (2002a) evaluated and designed SBR systems for simultaneous biological nitrogen and phosphorus removal for wastewaters from small communities subject to seasonal fluctuations in quality and quantity. A fundamental relationship between the effective sludge age, the recycle ratio and the cycle time was developed and used for the determination of physical design parameters. Artan *et al.* (2003) defined a rational design approach for organic carbon and nitrogen removal by SBR treating high-strength industrial wastewaters. Artan *et al.* (2004) structured a conceptual approach and a design methodology comparing a continuous-flow activated sludge and an SBR, both operated in a pre-denitrification mode, treating strong wastewater for nitrogen removal. The design approach was tested for two types of strong industrial wastewaters with different nitrogen contents and COD/N ratios, and a typical domestic sewage for comparative evaluation.

In all these studies, the significance of process modelling was underlined and model verification was used as an integral complement in support and verification of mass balance. Brenner (2000) used a mathematical model describing the fate of nitrogen and phosphorus in an SBR treating domestic wastewater and suggested modelling for the improvement of process analysis, design and control. Artan *et al.* (2002b) evaluated SBR performance for simultaneous nitrogen and phosphorus removal, covering all major process configurations with different aerobic, anoxic, anaerobic sequences and filling patterns. They established basic relationships between modelling and design based on overall process stoichiometry for the interpretation of the nutrient profiles obtained for different modes of operation. A similar approach was also used for the assessment of the effect of major operating parameters on system performance. These studies highlighted the fact that model simulation of SBR performance provides useful and reliable information for a selected set of different operating conditions. However, interpretation of the simulation results for process design is only meaningful when support is provided in terms of relevant process stoichiometry and mass balance relationships for model components.

1.3.2. Objective and Scope

Until recently, the design and operation of the SBR plants relied largely upon empirical knowledge. Such techniques, coupled with common sense and ingenuity, were generally successful for the specific cases where they were utilized. Specific cyclic operation schemes were arbitrarily selected to suit the needs of a particular case, mostly as an engineering option and not because they represented the most suitable operation alternative from a process standpoint. It is now commonly agreed that the only way to provide a reliable basis for design and operation is to identify a rational mechanistic description of the process in terms of microbial kinetics and mass balance.

This report has been prepared with the main objective to provide a unified basis for a rational design approach for SBR systems, primarily based on relevant process stoichiometry. Specific emphasis has been placed upon the fact that such a unified design approach is also by nature the determining factor for the selection of the most appropriate cyclic operation scheme, the sequence of necessary phases and filling patterns for the particular application. The proposed basis for design is developed and presented in a stepwise approach to cover both organic carbon and nutrient removal, domestic and industrial wastewaters, strong and specific wastes. The merit of model simulation as an integral complement of process design is also emphasized.

The report only covers suspended-growth SBR systems, as the intricate array of different microbial mechanisms taking place in biofilm systems are yet to be fully understood for being incorporated into similar design methodologies.

In the report, a general description of the process, in terms of mechanistic definition of major parameters and process options involved is presented in Chapter 2.. Process design for organic carbon, including design for strong industrial wastewaters, is outlined in Chapter 3. Chapter 4 discusses the basic principles and a unified design procedure for nitrogen removal and covers different aspects of phosphorus removal. Retrofitting of existing SBR systems for nutrient removal is also included in this chapter. Performance evaluation of SBR systems by model simulation is illustrated in the next Chapter.

2

Process Description

2.1. GENERAL DESCRIPTION

A general description of the SBR process, in a way that may be readily utilized in relevant stoichiometric relationships and mass balance equations, can best be made by identification and appropriate mechanistic definition of the set of operation parameters reflecting basic differences of the process with the continuous-flow activated sludge system.

The SBR process basically consists of a single tank, serving both as a biological reactor and settler: in SBR systems, biological processes and settling are carried out in a temporal sequence within the same reactor, whereas they occur simultaneously but in different tanks in continuous-flow systems. Consequently a portion of the reactor operation is considered to be inactive in terms of the microbial processes involved.

The SBR process incorporates a variable-volume reactor. The total reactor volume, V_T, is composed of two independently controllable fractions, namely a stationary volume V_0, that basically holds the settled sludge and a volume V_F which is the volume of wastewater that is filled and discharged every cycle (Figure 2.1). Therefore, the reactor volume reaches its maximum level at the end of the influent filling period.

The process involves a cyclic operation at steady-state with intermittent feeding during selected portions or the entire duration of the cycle. The duration of a cycle and the feeding pattern within the cycle can be independently adjusted for a selected mode of operation.

In this context, the definition of the SBR system, as an activated sludge process involving variable-volume reactors with a cyclic operation and each cycle having a fixed pattern of several

 Mechanism and Design of Sequencing Batch Reactors for Nutrient Removal: Scientific and Technical Report No. 19 by Nazik Artan and Derin Orhon. Published by IWA Publishing, London, UK. ISBN: 1843390825.

phases in succession, inevitably includes a set of basic parameters inherently associated with system operation. These parameters are identified and briefly described in the next section.

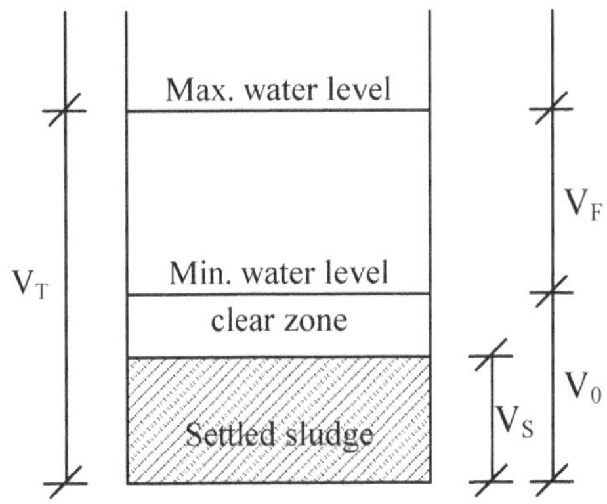

Figure 2.1. Schematic representation of SBR.

2.1.1. Cycle Frequency (m)

Number of cycles per day, m, is an important parameter to be selected in SBR design and operation. It defines the total cycle time, T_C and the fill volume per cycle, V_F:

$$T_C = \frac{1}{m} \tag{2.1}$$

$$V_F = \frac{Q}{m} \tag{2.2}$$

where Q is the daily volumetric flow rate of the wastewater to be treated in the SBR system.

2.1.2. Nominal Hydraulic Retention Time (HRT)

A nominal hydraulic retention time, θ_h is usually defined for SBRs similar to continuous-flow activated sludge systems as V_T/Q and can be expressed as a function of T_C using above equations:

$$\theta_h = \frac{V_T}{Q} = \frac{V_0 + V_F}{mV_F} = \left(1 + \frac{V_0}{V_F}\right) T_C \tag{2.3}$$

It should be noted that V_0/V_F ratio corresponds to the recycle ratio of the system. Similarly, θ_h may also be expressed in terms of the volumetric exchange ratio, VER, which is defined as the ratio of the fill volume to the total volume.

$$\theta_h = \frac{V_T}{mV_F} = \frac{T_C}{VER} \tag{2.4}$$

2.1.3. Duration of Phases in a Cycle

The SBR is a biological treatment process with an inherent cyclic operation, each cycle incorporating the same pattern of several successive phases. Traditionally, a cycle in the SBR process is conceived to consist of five phases: *fill, react, settle, draw* and *idle* (Irvine and Bush, 1979). In the *fill phase* (T_F), influent wastewater is fed into the SBR reactor, starting the relevant biochemical reactions through mixing/aeration as necessary, with the exception of static filling. In the *react phase* (T_R), influent filling stops and the SBR is operated as a batch reactor with mixing/aeration continuing until reaching the desired level of completion for targeted biochemical conversions. Mixing/aeration is discontinued at the beginning of the *settle phase* (T_S), and the sludge is allowed to settle to the reactor bottom throughout the phase.. The clear supernatant over the settled sludge layer is decanted and discharged during the *draw phase* (T_D).The *idle phase* (T_I) serves as a reserve period intended to increase the overall operational flexibility of the system. It can be used to extend the duration of one or more of the other phases when need arises.

In today's practice, the duration of wastewater feeding to the SBR may be varied from an almost instantaneous dump fill, mostly for creating create substrate accumulation or sustaining selector conditions, to a continuous fill persisting throughout the entire process cycle including the settle and draw phases, making *react* period unidentifiable as a discrete process phase. In this context, wastewater filling, (T_F), provided during the desired portions of the cycle, should not be considered as a mechanistic component of the cycle, mainly because it does not necessarily precede a distinct react phase and biological reactions obviously occur during wastewater filling, which only affects the structure of the relevant mass balance equations. Therefore, in lieu of fill and react phases, adoption of the concept of a *process phase* (T_P), covering the whole period adjusted to sustain the selected biochemical reactions in the reactor, seems more appropriate from a process standpoint. As the cycle includes, as mentioned before, an additional period (T_S+T_D+T_I) assumedly with no biological conversion for simplicity of mechanistic evaluation and modelling, the total cycle time, T_C, can be defined as:

$$T_C = T_P + T_{S+D+I} \tag{2.5}$$

The duration of fill time, T_F, can range from a small fraction of total cycle time to total process time, or even to the total cycle time and fill time ratio, FTR., is defined as

$$FTR = T_F / T_C \tag{2.6}$$

Fill time ratio affects reactor hydraulics seriously. The lower the fill time ratio the more pronounced becomes the substrate gradient within a cycle.

2.1.4. Duration of Periods in a Process Phase

The process phase can be fully aerobic as in the systems for COD removal or can have various environmental conditions adjusted by energy input. In nutrient removal SBR systems, the process phase T_P consists of aerated periods, (T_A), and mixed periods, (T_M), which can be anoxic or anaerobic depending on the presence of nitrate:

$$T_P = T_M + T_A = T_{AN} + T_{AOX} + T_A \tag{2.7}$$

2.1.5. Number of Tanks

The SBR process can be carried out in a single reactor or in multiple reactors in parallel. If equalization is not used, a continuous wastewater flow can be accommodated by providing multiple reactors; for n reactors, the equation below must be observed.

$$T_C = nT_F \tag{2.8}$$

Thus, n is inversely proportional to the FTR:

$$FTR = \frac{1}{n} \tag{2.9}$$

2.1.6. Sludge Retention Time (SRT)

The sludge retention time (SRT) for SBR system, θ_X, is defined as the mass of sludge contained in the reactor, M_X, divided by the sludge wasted per day, P_{XT}, as for the continuous-flow systems.

$$\theta_X = \frac{M_X}{P_{XT}} \tag{2.10}$$

Sludge wasting from the mixed liquor (i.e. before settling) may provide a simple and direct means of controlling θ_X. In this case, θ_X may be defined, as follows, in terms of the total reactor volume, V_T, the volume of the sludge wasted from mixed liquor each cycle, V_W, and the cycle time, T_C. In practice however, excess sludge wasting takes place mostly after settle phase on a daily or even weekly basis.

$$\theta_X = \frac{V_T X_T}{m\, V_W X_T} = \frac{V_T}{V_W} T_C \tag{2.11}$$

The aerobic sludge age, θ_{XA}, is defined as a function of the total aerated periods within the cycle, the only period of the operating cycle sustaining autotrophic growth:

$$\theta_{XA} = \theta_X \frac{T_A}{T_C} \tag{2.12}$$

Heterotrophic growth can take place during both the aerobic and anoxic periods. Growth and endogenous respiration processes are assumed to cease during the anaerobic period. Hence, the effective period, T_E, is the sum of the aerobic, T_A, and anoxic, T_{AOX}, periods in nutrient removal systems while it is totally aerobic in carbon removal and nitrification systems. Thus, effective sludge retention time for heterotrophs is defined as:

$$\theta_{XE} = \theta_X \frac{T_E}{T_C} = \theta_X \frac{T_A + T_{AOX}}{T_C} \tag{2.13}$$

The relationship between aerobic and effective sludge retention times can be expressed as below using above equations:

$$\theta_{XE} = \frac{\theta_{XA}}{\left(1 - T_{AOX}/T_E\right)} \tag{2.14}$$

2.2. REACTOR HYDRAULICS

Interpretation and control of the efficiency of the SBR systems requires first a proper kinetic description of the reactions taking place in the reactor. *Reaction kinetics* however can not be directly evaluated and interpreted as process performance without any consideration of the *reactor hydraulics* defining the mixing characteristics of reactors where biological reactions are allowed to occur. In order to obtain accurate information about the fate of any component in the reactor, the conversion rate of the specific component should be processed as dictated by reactor hydraulics in terms of appropriate mass balances around the system.

On the basis of its hydraulic characteristics, SBR may be considered as a batch reactor during the portion of the cycle with no without filling. This analogy however is not accurate and may at times be misleading for two main reasons: (i) Actually, SBR is a *continuous* activated sludge system that reaches, if properly operated, a cyclic steady-state for a given set of operating conditions such as feed composition, structure of the cycle, etc. The term "batch" in the name of the process only implies intermittent feeding and discharge and nothing else. At steady state, the progression of each cycle is identical and the overall behaviour of the system in terms of concentration transients of significant parameters within the cycle remains the same from one cycle to the following one. The necessary period for a SBR system to reach steady-state is variable but at least two sludge ages. (ii) SBR sustains, as similar continuous-flow activated sludge systems, a selected *enriched* microbial culture and the *enriched culture* that prevails at steady-state reflects the specific characteristics of the selected operating conditions. It is then obvious that for every specific SBR operation, the nature and composition of the enriched culture will be different. In an excellent review on this subject, Grady *et al.,* (1996) underline the importance of *culture history* and state that a batch reactor develops a microbial growth medium totally different from conditions associated with activated sludge growing in continuous systems.

As previously mentioned, wastewater feeding in SBR operation may take place during the selected fractions of the cycle but effluent withdrawal occurs periodically so that the system needs to be evaluated as a *variable volume* reactor. During the fill phase, the volume of the each tank operated in parallel gradually increases from an initial level, V_0, to the total reactor volume, V_T, defined as

$$V_T = V_0 + Q_{in}\,T_F \tag{2.15}$$

where Q_{in} is the volumetric flow rate of the influent to the one of the SBR.

The mass balance for any model component present in the influent, C_A, may be formulated as follows, taking the hole reactor volume as the control volume, with the assumption that a SBR always operates as a completely mixed reactor,

$$\frac{d[VC_A]}{dt} = Q_{in} C_{A\,in} + Vr_A \tag{2.16}$$

where:

V = reactor volume at a given time, t

$C_{A\,in}$ = influent concentration

r_A = observed conversion rate for C_A

Taking the differential of the left-hand side and rearranging to include the definition of flow rate, dV/dt=Q, Eq 2.16 may be rewritten as:

$$\frac{dC_A}{dt} = \frac{Q_{in}}{V_{0i} + Q_{in}\,t}\left(C_{Ain} - C_A\right) + r_A \tag{2.17}$$

where C_A is the reactant concentration at a given time during the process phase. During the *react* portion of the process phase (i.e. after fill phase) where the SBR functions as a completely mixed batch reactor having constant volume, mass balance equations are reduced to only the reaction terms:

$$\frac{dC_A}{dt} = r_A \tag{2.18}$$

When distance is translated into the flow-through time, the above equation and the equation which describes the fate of a model component in a plug-flow reactor (PFR) are identical. They yield identical concentration profiles if they are integrated for the same concentration limits.

With the commonly accepted assumption that no biological conversion takes place during settling, draw and idle phases, the concentration of a soluble component at the end of the process phase remains the same till the beginning of the next cycle. When the system reaches a cyclic steady-state through continuous operation with constant influent conditions, the initial concentration of a soluble component, S_{ini}, becomes equal to the concentration of the same component at the end of the cycle i.e. the effluent concentration, S_e. The mass and concentration of a particulate component, however, is bound to change due to drawing of supernatant and sludge wasting according to the selected SRT to maintain cyclic steady-state operation. Assuming that excess sludge is wasted at the end of the react phase at the rate dictated by the selected sludge age, θ_X, and settling is ideal with no carry-over of suspended solids in the effluent stream, the concentration of any particulate component at the beginning of the next cycle, X_{ini}, can be calculated from the following mass balance expression:

$$V_0 X_{ini} = V_T X_n \left(1 - 1/m\theta_X\right) \tag{2.19}$$

where X_n is the concentration of any particulate component at the end of the process phase. A selected biokinetic model can be implemented in various computer codes which use mostly numerical integration methods to simulate the behaviour of an SBR as mass balance expressions lead to nonlinear equations with no analytical solution. Using the calculated S_{ini} and X_{ini} concentrations as initial values, the system will reach the steady state after several SRTs. Therefore, the SBR performance at steady state operated with very low FTRs (i.e. dump filling) may be assumed to approximate that of the PFR activated sludge configuration in continuous-flow systems keeping in mind that ideal plug-flow conditions can not be achieved in an aeration basin. Multiple completely mixed flow, CMF, reactors in series models are often used to describe non-ideal flow which is a far more convenient approach, especially for steady state modelling. The steady state response of a cascade of infinitesimal CMF reactors is identical to that of both the PF and batch reactors (Figure 2.2.(a)) The recycle ratio in the continuous-flow system is comparable to the V_0/V_F ratio of SBR and cycle time corresponds to the flow-through time, τ, given by θ / (1+R). Arguments are also presented in the literature to compare the operation of a SBR with a slow fill to a single CMF reactor (Irvine and Ketchum, 1989). This type of analogy does not generally hold true as discussed in Chapter 5. In fact, SBR with slow fill or extended fill without *react phase* resembles CMF reactors in series using step-feed operation as illustrated in Figure 2.2(b).

The adopted operation mode may exert a significant impact on the hydraulic characteristics and on the resulting concentration profiles of model components within a cycle. The fill time ratio, FTR and the recycle ratio, V_0/V_F, should be considered as the important parameters having significant effect on the resulting concentration profiles. However, filling pattern exerts no significant effect on the steady state effluent COD for a given effective SRT for the treatment of a domestic wastewater having low fraction of soluble readily biodegradable COD.

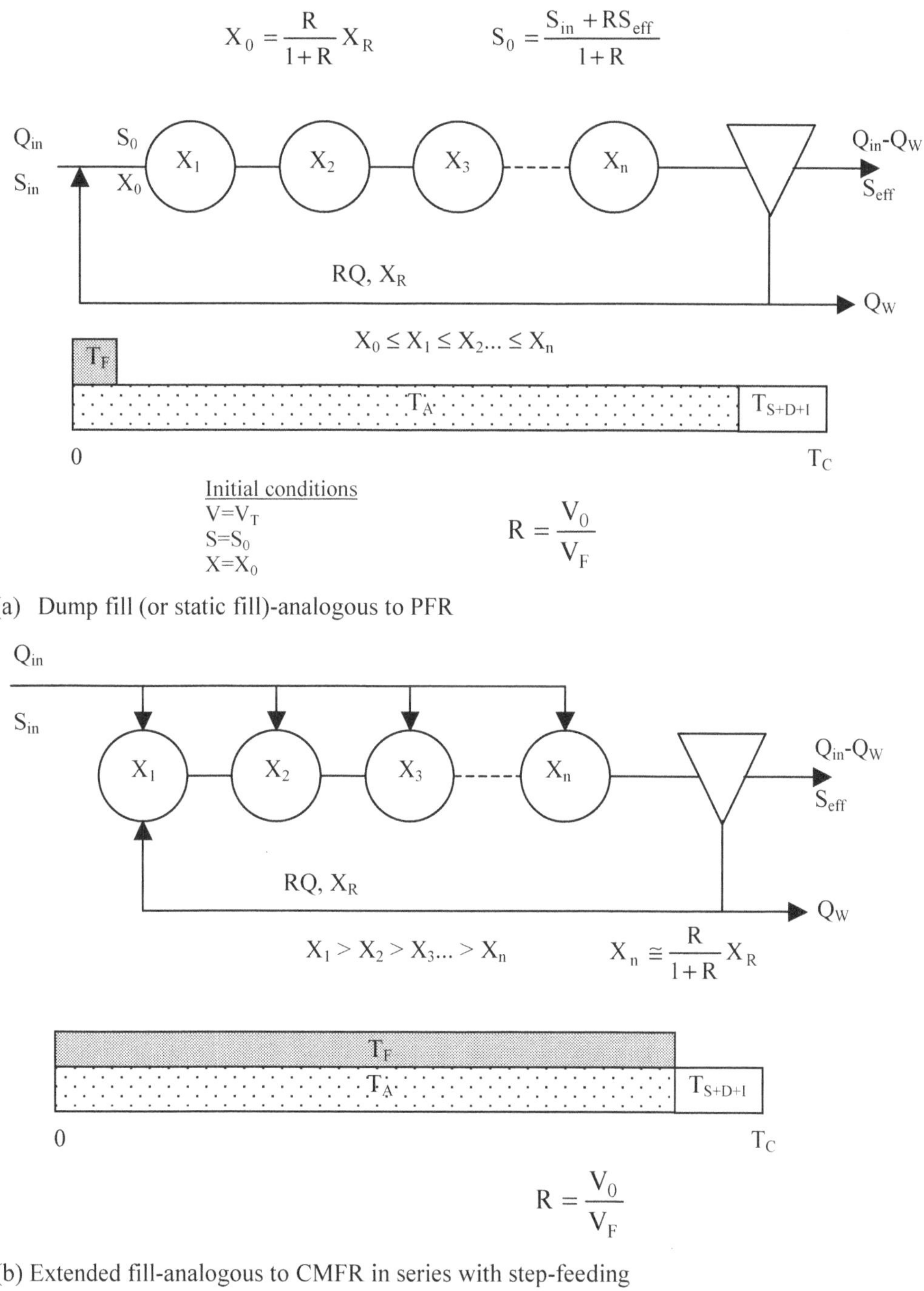

Figure 2.2. Comparison between SBR and continuous flow systems

On the other hand, in the case of industrial wastewater treatment or for nitrification process, filling pattern may have important effect on the effluent quality. This will be illustrated with the simulations given in Chapter 5 and has important implications. Excluding oxygen demand distribution, important issues like total oxygen requirement, excess biomass production rate and hence total biomass in the reactor can be determined by process stoichiometry regardless of the reactor hydraulics once the effective SRT has been fixed. This means that expressions derived for a single CMFR operating at a steady state, which can be solved analytically, can be used to estimate the required oxygen and biomass for an SBR system. The other effects of filling pattern will be discussed in the ensuing sections.

2.3. PROCESS OPTIONS

SBR technology has been successfully used for both carbon and nutrient removal. The flexibility of operation has always been the main asset of the SBR; simple manipulation of operating parameters has been extensively explored and practiced for the generation of a wide range of process options, especially for nutrient removal. Achievement of the desired system performance depends mainly upon the appropriate sequence and duration of different phases and periods within the cycle, aside from the right choice of design parameters such as the sludge age as in continuous-flow systems. In this framework, selection of the suitable process option for a targeted operation must then rely upon and fully benefit from the necessary information on the basic principles of the type of microorganisms and related biochemical reactions involved In this context, different SBR process options may be tailored specifically for the particular effluent discharge limitation applicable for each case.

2.3.1. Carbon Removal and Nitrification

There are no different process options for the SBR systems that are used for carbon removal alone and combined with nitrification in terms of the aeration pattern since they both rely on aerobic processes. Accordingly, organic carbon removal and nitrification may take place in the same reactor with a single aerobic process phase within cyclic operation. The particular SBR operation may only be different according to the filling strategy adopted and the measures used to handle periodic influent and variable flow conditions.

Two or more tanks are normally used for handling periodic influent. In practice, increase in the number of reactors is very likely to cause an increase in the total equipment cost. For larger plants, it may be practical to have more than two tanks since it allows a higher degree of flexibility to handle variable flow conditions. An influent holding tank can also be incorporated to the system in order to maintain control of the fill rate. SBR systems that operate with a continuous wastewater feeding during the cycle use special measures (i.e. separation with a baffle) to maintain quiescent settling and minimize contamination of the treated effluent by the inflowing wastewater during settle and draw phases. The three types of filling strategies used for carbon removal are illustrated in Fig 2.3. Similar options with different aeration patterns are also used for nutrient removal processes.

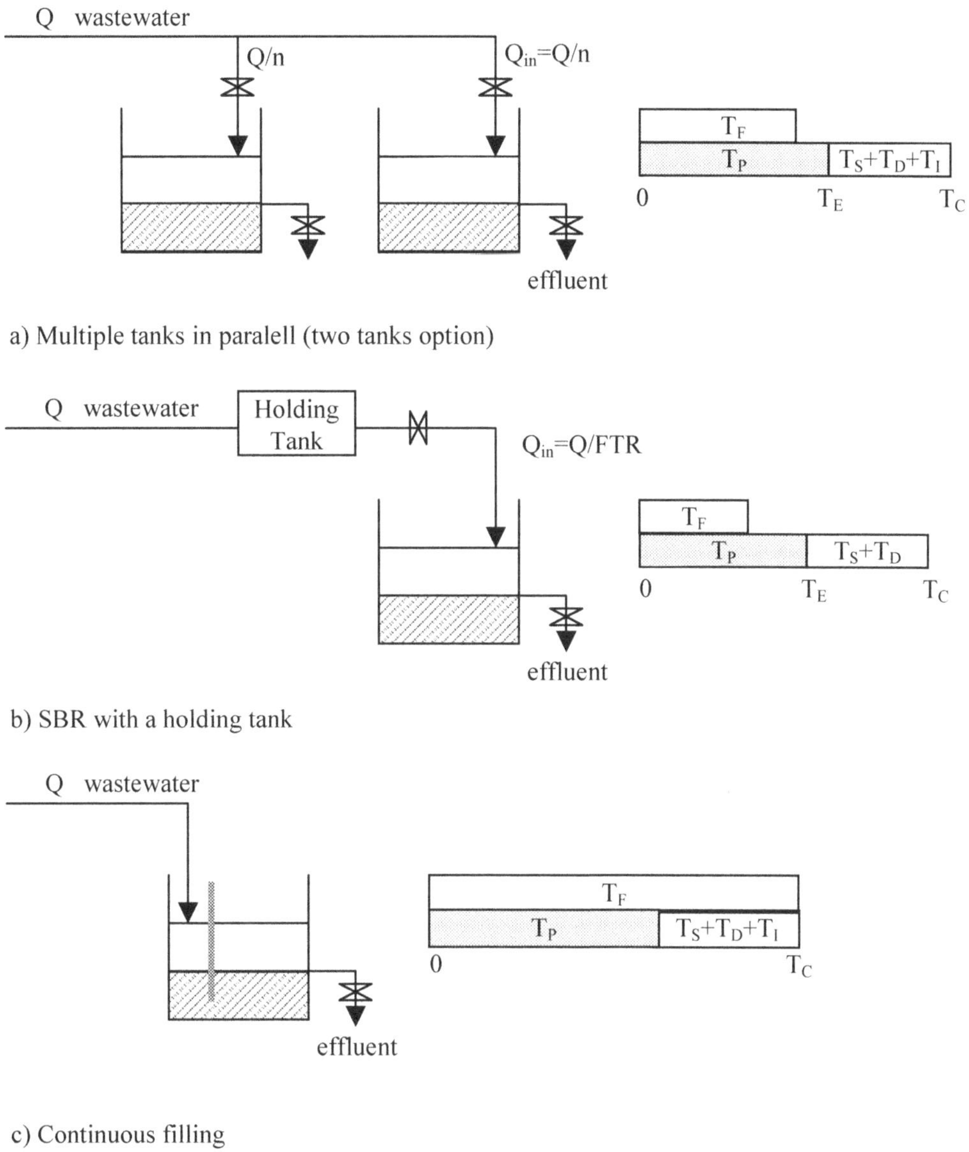

Figure 2.3. Three types of filling strategies

2.3.2. Nutrient Removal

Nutrient removal in SBR can be accomplished with relative ease when appropriate conditions for nitrification, denitrification and Enhanced Biological Phosphate Removal (EBPR) are created. The SBR offers a great deal of operational flexibility as it allows for easy adjustment of aerobic, anoxic and anaerobic periods through temporal control of mixing/aeration and filling. Most of the advantages of the SBR are attributed to the very flexible nature of the operating parameters. However, the system flexibility of SBR for nutrient removal is seriously challenged because nitrate recycle (R_N) cannot be handled and controlled separately from sludge recycle (R_S) as a result of temporal sequence. This is a major limitation for the SBR compared to continuous-flow activated sludge which offers independent control of internal recycle. Furthermore, the advantages may also be interpreted as handicaps if the flexibility in design and operation complicates the basic mechanistic understanding of the process, a crucial issue for successful application.

Nutrient removal process configurations may be categorized according to their nutrient removal capabilities as: i) nitrogen removal processes ii) phosphorus removal processes and iii) simultaneous nitrogen and phosphorus removal processes

2.3.2.1. Nitrogen Removal

Three types of aeration and filling strategies may be envisaged for nitrogen removal, as illustrated in Figure 2.3. The first one employs a single mixed/non-aerated period followed by an aerobic period together with wastewater feeding within the mixed phase (Figure 2.4.(a)). Anoxic conditions are obtained through nitrate recycle from the previous cycle to the mixed period. This operation mode as a pre-denitrification system is comparable to the modified Ludzack-Ettinger process, (MLE), in continuous-flow activated sludge system with the understanding that V_0 /V_F ratio is equal to the total recycle ratio of the MLE process. In the MLE process both R_S and R_N are recycled together to the mixed/anoxic reactor, because there is no need to separately handle the two recycle streams in terms of the biochemical processes involved. Therefore, the MLE process can be duplicated with temporal sequence associated with the SBR operation.

In all predenitrification systems the level of denitrification, hence the degree of nitrogen removal efficiency is limited by the recycle ratio. A post-denitrification step is usually introduced to the process scheme to overcome the recycle limitation. In continuous-flow systems, the *4-stage Bardenpho process* for example, uses post-denitrification as a polishing step after a high degree of pre-denitrification so that it can secure almost complete denitrification depending on suitable wastewater characteristics. In this process, nitrate to be introduced to the first anoxic reactor through sludge recycle is negligible. The available nitrate for pre-denitrification the first anoxic reactor is supplied by internal nitrate recycle from the first aerobic reactor which also serves as the nitrate pool for the following second anoxic reactor. As R_S and R_N are recycled to the first aerobic reactor from different steps, the 4-stage Bardenpho process cannot be mimicked in a temporal sequence. Consequently, an SBR operation with a dual anoxic/aerobic phase sequence within a cycle (Figure 2.4.(b)) should not be interpreted as the equivalent of a 4-stage Bardenpho process and complete denitrification cannot be expected from this process. On the contrary, two or three step mixing/aeration operation mode in SBR is comparable with the *step-feeding,* (SF), activated sludge process in continuous-flow systems. In this operation mode the SBR should be filled during the mixed periods in every sub-cycle for effective use of the carbon source in the wastewater.

More frequent shifts in aeration (intermittent aeration) with a multiple sequence of anoxic/aerobic phases may be applied as required. In this case, it may not be practical to restrict wastewater filling to only anoxic phases and filling may be adjusted to continue during the entire process phase (*extended filling*). This filling and aeration strategy (Figure 2.4.(c)), called the *IAEF process* (*intermittent aeration extended filling)*, may be evaluated in a way similar to the *intermittent aeration,* (IA), configuration of continuous-flow activated sludge systems, keeping in mind that the IA process has a continuous effluent with a periodic variation in the concentration of nitrogen species whereas in the IAEF mode of SBR operation the concentrations in the effluent are constant but flow is intermittent.

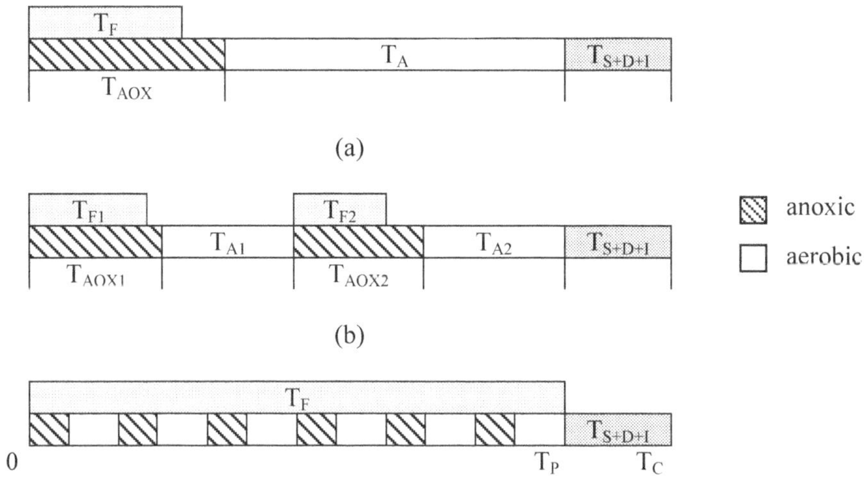

Figure 2.4. SBR configurations for nitrogen removal (a) pre-denitrification; (b) dual anoxic filling; (c) Extended filling with multiple anoxic periods

Simultaneous nitrification–denitrification may be observed in SBR operation, especially in systems aerated with surface aerators. In this type of operation, it is difficult to identify distinct aerobic and anoxic periods within the process phase. Nitrification and denitrification may alternately dominate depending on the fluctuations in the dissolved oxygen concentration profile sustained in the reactor. This way, the process phase can be effectively used both for nitrification and denitrification for an average dissolved oxygen concentration of around 1.0 mg/l in the mixed liquor. Consequently, the V_0 /V_F ratio does not exert a decisive impact on the effluent nitrate concentration in these systems

2.3.2.2. Phosphorus Removal

Enhanced biological phosphorus removal, (EBPR), occurs with the predominance of a group of bacteria, commonly called phosphate-accumulating organisms, (PAOs), capable of storing polyphosphate within the cell. EBPR is only sustained when the activated sludge operation involves an anaerobic/aerobic sequence. When nitrogen removal is not required or desired, a two-stage process, called the *A/O Process* provides the simplest flow scheme required for EBPR in continuous-flow activated sludge systems. The A/O Process is also applicable in the SBR cycle. In this system, the process phase incorporates a sequence of a mixed period without aeration for anaerobic microbial activity followed by an aerobic period (Figure 2.5.(a)). The aerobic sludge age is selected below the minimum SRT for nitrifiers. In cases when unintended nitrification takes place, the V_0 /V_F ratio should be selected as low as possible to decrease the intrusion of nitrate into the mixed period. When seasonal or year-round nitrification occurs, mixing with no wastewater filling can be implemented during the idle phase so that the level of nitrate recycle is reduced by endogenous respiration. This additional phase may help to minimize or avoid the anoxic phase that would be established within the mixed period through contact of nitrate containing recycle with the wastewater containing readily biodegradable substrate. The denitrification rate in the idle phase (without filling) may be relatively high due to the short SRT applied (Figure 2.5.(b)).

2.3.2.3 Simultaneous Nitrogen and Phosphorus Removal

Process configurations for SBR for simultaneous nitrogen and phosphorus require special care in design and operation. Long SRTs must be selected for efficient nitrogen removal. On the other hand, EBPR may be seriously affected by denitrification efficiency since nitrate will be recycled with the sludge at the end of the cycle. An SBR operated for nitrogen removal is not capable of starting the process sequence with an anaerobic phase. The stationary volume, V_0 becomes the nitrate pool generated through nitrification in the aerobic phase and at the start of the following cycle, this pool inevitably consumes the biodegradable substrate in the feed stream. Consequently, a pre-anoxic period is inherently established at the beginning of the cycle and the temporal sequence in the SBR does not allow for an anerobic period followed by an anoxic period as in the continuous-flow nutrient removal activated sludge configurations such as the *A2O*, *5-stage Bardenpho* or *UCT* processes. It is possible to shorten the pre-anoxic period, however, the second constraint may be more important for the efficient use of carbon source, in view of the fact that PAOs can also denitrify. The anoxic-anaerobic-aerobic sequence in the SBR cycle (Figure 2.5.(c)) will lead to the consumption of organic carbon stored by PAOs under anaerobic conditions in the subsequent aerobic period instead of being used for denitrification. If a low V_0 /V_F ratio is selected to sustain anaerobic conditions after a short pre-anoxic period, nitrogen removal by predenitrification will be limited. Further nitrogen removal based on post-denitrification may be obtained using a second set of mixed/aerated periods (Figure 2.5.(d)). Since internal recycle is impossible, aerobic conditions need to be established prior to the second anoxic period to produce available nitrate for post-denitrification. Filling time should not be extended to the aerated periods since internal carbon source will likely be limited to satisfy both nitrogen removal and EBPR. Fast filling strategy that leads to high substrate concentration may also help to trigger and activate the storage mechanism, and in turn a good EBPR.

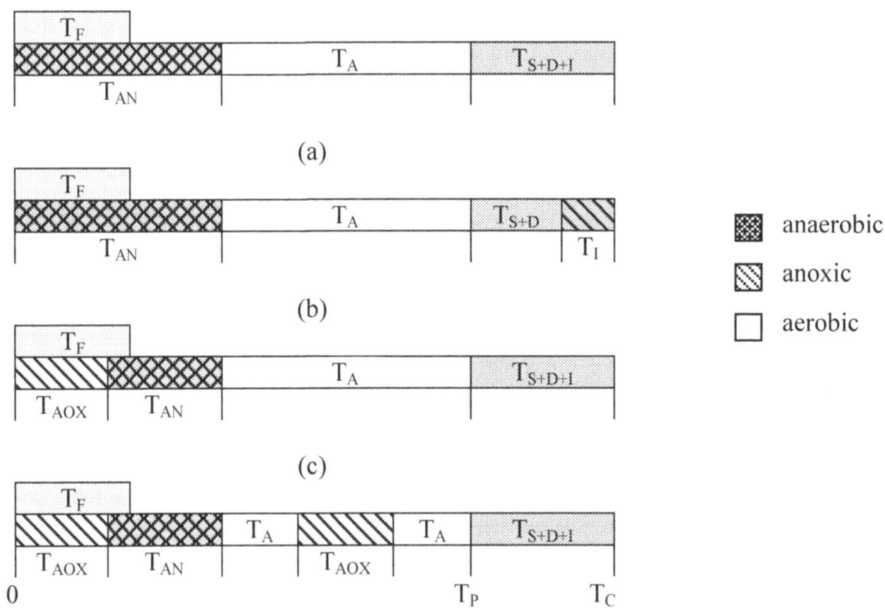

Figure 2.5. SBR configurations for (a) EBPR only; (b) EBPR with seasonal nitrification (c) EBPR with partial nitrogen removal; (d) joint nitrogen and phosphorus removal.

2.3.2.4 Nutrient Removal with External Nitrification

External nitrification in a two-sludge system can be a viable alternative for simultaneous N&P removal. In this process EBPR and denitrification processes are carried out in the main SBR whereas nitrification is allowed to proceed in another reactor, specifically suitable for this purpose, such as the *Sequencing Batch Biofilm Reactor,* (SBBR). The main virtue of this process configuration is the capability of altering the inherent deficiency of the SBR process in establishing an anaerobic/anoxic sequence within the cycle and exercising independent control of the anaerobic, anoxic and aerobic periods. This way, the main SBR may essentially be operated with an anaerobic/anoxic sequence during the process phase as in similar continuous-flow activated sludge configurations. Anoxic conditions can be independently sustained via filling with nitrified effluent from the external nitrification reactor (SBBR).

The basic operation mode of this SBR system is schematically illustrated in Figure 2.6. Wastewater is fed into the main SBR onto the settled activated sludge left from the previous cycle. Anaerobic conditions are maintained to store the organic carbon as PHA by PAOs. At this point, mixing is stopped allowing the sludge to settle and the ammonium/phosphate rich clear supernatant - due to the P release under anaerobic conditions - to be replaced by the nitrified effluent of the external nitrification reactor. The volumetric exchange ratio, VER, (the ratio of the replacement volume to the total reactor volume) should be as high as possible to reduce the amount of ammonium-N remaining in the main SBR at the end of this portion of the cycle. It is now well established that PAOs can grow on the stored substrate with the uptake of phosphate under anoxic conditions. This way, the anaerobic/anoxic sequence implemented in the cycle allows the use of the carbon source in the influent both for growth of PAOs and denitrification. The main SBR is operated at relatively low SRT levels since there is no need for nitrification in this reactor. A short aerobic period should be employed for re-aeration immediately before the settling phase. This aerobic period may be necessary for P uptake if nitrate is consumed before phosphate. Considering that the effluent of the external nitrification reactor contains nitrifiers, this aerobic period may be adequate for nitrification in the main reactor. Multiple tanks operated in parallel and consistent in number with the selected FTR can be used as the main SBR.

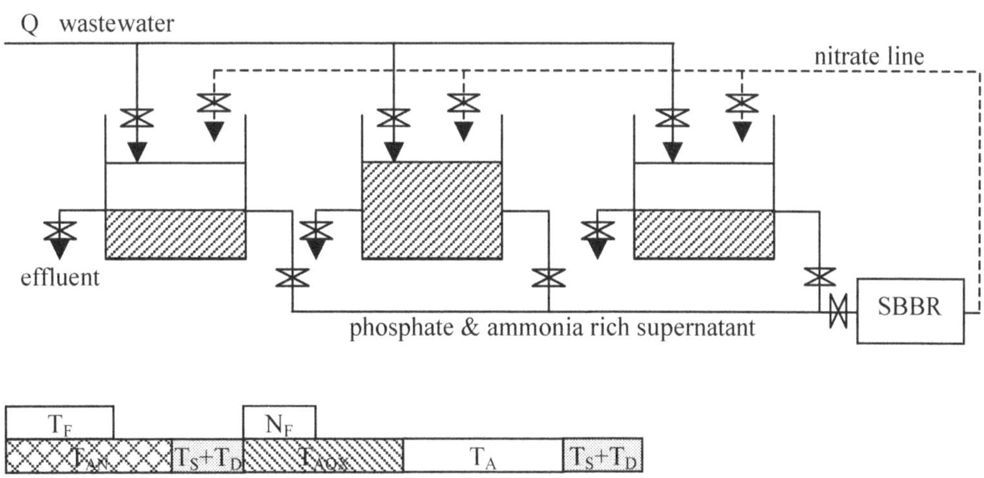

Figure 2.6. Nitrogen and Phosphorus Removal using External Nitrification.

The external nitrification reactor can be operated with a high SRT using a short cycle period since the influent biodegradable COD level is very low. The fill volume of this reactor must, for practical purposes be selected equal to the fill volume of the one of the parallel SBRs. However the external nitrification reactor can serve all SBRs as the period of its cyclic operation will be substantially shorter than the SBR cycle (Fig 2.6).

2.4 EFFECTS OF FILLING AND AERATION PATTERNS ON POPULATION DYNAMICS

The filling pattern may also influence the sludge settling characteristics especially in carbon removal systems through selection of desirable floc-forming or filamentous microorganisms. The selection mechanism, called kinetic selection, acts by creating conditions under which the growth rate is significantly higher for floc-formers than for filaments. According to the saturation kinetics model, the high substrate concentration under aerobic conditions in the beginning of the cycle will result in favourable conditions for the dominance of floc-formers having high maximum specific growth rate. In SBRs the actual substrate concentration is closely related to the filling rate. The higher the filling rate, the more pronounced the substrate gradient becomes within a cycle as explained in the previous section. It is reported that there may be also a competitive disadvantage for some filamentous bacteria having high growth rates at high substrate concentrations and comparatively high decay rates in the absence of substrate when feast/famine conditions are imposed.

There is evidence that the introduction of anoxic conditions always resulted in very fast and effective improvement of sludge settling characteristics. Based on the observation that the specific denitrification rate of bulking sludge is lower than that of well-settling sludges by an order of magnitude it was claimed that filamentous microorganisms were not able to use nitrate as an electron acceptor. Therefore, anoxic conditions cause blocking of the metabolic pathways of substrate utilization of filaments and suppress their growth. This method of microbial selection is called metabolic. Two characteristic features of metabolic selection are the extremely fast response and complete elimination of filaments from activated sludge. Metabolic selective pressure prevailing in nutrient removal systems plays a far more important role than kinetic selection in controlling filamentous bulking (Wanner 1994).

On the other hand, it is pointed out that too high anoxic (actually unaerated) fraction may lead to deterioration of sludge settling characteristics although anaerobic-anoxic SBRs can be operated by some researchers. It is also indicated that anoxic conditions significantly intensified selective pressure resulting in the disappearance of filaments and abundant growth of viscous zoogleal colonies producing excessive extracellular polymeric substances (EPS), which eventually causes non-filamentous bulking. Because of the importance of producing suitable biomass structure that allows easy separation from the treated effluent, further research on this topic is required (Artan and Tasli, 1999).

In nutrient removal SBR, the filling pattern plays a significant part in efficient utilization of the internal carbon rather than in creating the concentration gradient needed for kinetic selection. There is a strong competition among phosphorus removing, denitrifying and aerobic heterotrophs for the readily biodegradable substrates since they are utilized with extremely high rates under all anaerobic, anoxic and aerobic conditions. Into which period these substrates are supplied decisively influences the microbial composition of activated sludge thus influencing nitrogen and phosphorus removal efficiency.

3

Process Design for Carbon Removal

3.1. BASIC PRINCIPLES

One of the most significant achievements in wastewater treatment is the current mechanistic interpretation of the activated sludge process. This interpretation has already been successfully translated into process kinetics, stoichiometry and applicable mass balances for the design of continuous-flow systems. This rational approach is somewhat more delicate and difficult for SBR processes incorporating a higher flexibility of operation and therefore, a higher level of options for selecting the optimum design strategy, It can only be accomplished if relevant principles of process kinetics and stoichiometry are properly integrated into a logically structured design procedure. In this context, a unified basis of a systematic design approach for different SBR configurations can be formulated in a way quite similar to that currently applicable for continuous-flow activated sludge, provided that it accurately accounts for specific characteristics and functions of the SBR. In general, a steady-state design model may be adopted since a long-term SBR operation under constant influent and operation conditions results in identical cycle profiles and effluent quality, which can be interpreted as steady-state conditions.

The merit of the design also relies upon the extent of relevant information available for appropriate evaluation. This information is specific for the wastewater to be treated and involves (i) wastewater characterization and (ii) required stoichiometric and kinetic data. Reliable information on wastewater flow characteristics, with average and peak values, daily variations and seasonal or specific fluctuations is a prerequisite for a proper design. Assessment of the organic substrate is also essential but not so simple, as wastewaters contain a great variety of organics, which cannot be identified individually on a routine basis. Consequently, collective substrate parameters are traditionally used as indirect index values to characterize the overall organic content. In this context, BOD_5, although generally included in wastewater characterization should not be considered as a design parameter. It merely represents an unknown fraction of the total organic content, variable for each wastewater tested. As such, it cannot be used in mass balance equations related to organic mater utilization or oxygen consumption. COD, unlike BOD_5, may be used as a direct parameter to yield the stoichiometric equivalent of carbonaceous substrate, provided that its biodegradable fraction is determined. This fraction reflects the electron balance between substrate, biomass and the electron acceptor. In this context, wastewater characterization requires the assessment of major fractions of the wastewater total COD, (C_{T1}), at least as biodegradable COD (C_{S1}), inert soluble COD (S_{I1}) and inert particulate COD (X_{I1}). If required, C_{S1} can be further fractionated into readily biodegradable COD, (S_{S1}) and slowly biodegradable COD, (X_{S1}). Extensive information is available in the literature on the characterization of domestic sewage with emphasis on COD fractionation (Henze 1992; Orhon *et al.,* 1997; Rössie and Pretorius, 2001). Aside from relevant COD fractions, wastewater characterization should primarily cover all parameters prescribed in related effluent discharge regulations; it should also give indications for pre-treatment, if needed and supply the necessary information on the suspended solids content, nutrient balance (wastewater N and P contents; COD/N; COD/P ratios), pH, buffer capacity (alkalinity) etc.

Evaluation of organic carbon utilization involves a systematic analysis of the behaviour of heterotrophic microorganisms, preferentially using dissolved oxygen as the final electron acceptor. A rational design for organic carbon removal does not require elaborate modelling or model simulation, as an integral part of the procedure. It should rely however on COD mass balances associated with an adopted process modelling. The simplest biokinetic model should include as major processes growth on S_S and endogenous respiration. Slowly biodegradable COD is hydrolyzed to S_S prior to being used by heterotrophs for growth. Particulate inert COD, (X_E) is produced as part of the endogenous respiration. Effluent COD contains mostly soluble inert COD in the influent or produced in the system. Biodegradable COD concentration in the effluent can be neglected in well-designed systems. Under the assumption that all biodegradable COD is consumed, the kinetic parameters of the adopted model are not required except for the endogenous respiration rate coefficient, b_H, for heterotrophs.

Process stoichiometry is basically defined by means of the yield coefficient, Y_H, setting the necessary mass balance relationships for excess sludge generation, oxygen consumption, etc. On the basis of aerobic growth energetics, this coefficient assumes a constant value when expressed in terms of g cellCOD/cellCOD (e^-.equivalent biomass/e^-.equivalent substrate). Ekama and Marais (1984) reported a Y_H value of 0.66 g cell COD/g COD based on experiments to characterize domestic wastewaters. Henze *et al.* (1995) suggested a default Y_H value of 0.64 g cell COD/g COD for the Activated Sludge Model No.2, (ASM2). This value, generally adopted in modelling studies, is experimentally confirmed by Ubay Çokgör (1997) for Istanbul domestic sewage. The design procedure also requires the inert fraction of biomass, f_E, a coefficient defining the stoichiometry of particulate microbial products generation through endogenous respiration. This coefficient is commonly used with a default value of 0.2 for domestic wastewaters.

The kinetic and stoichiometric data are quite wastewater specific and should be experimentally determined in conformity with the structure of the model adopted. While the process kinetics and stoichiometry are well documented for domestic sewage, experimental support is still needed for most industrial wastewaters. A comprehensive survey on wastewater characterization, COD fractionation and stoichiometric/kinetic data for a wide range of industrial wastewaters is presented by Orhon *et al.* (1999).

3.2. SELECTION OF SRT

The sludge retention time, SRT, is the most significant parameter for design and operation since it is related to the average net specific growth rate of the system. Net average specific growth rate is defined as the mass of biomass produced per unit time, P_{XH}, divided by the biomass contained in the reactor using the mass balance for active biomass, X_H.

$$\mu_H - b_H = \frac{P_{XH}}{V_T X_H} \tag{3.1}$$

As the specific character of the SBR, biomass production can only take place during the effective time, T_E, since a portion of the cycle has to be devoted to settle and draw. On the other hand, at steady state the mass of sludge wasted must equal the net biomass produced. Hence:

$$P_{XH} T_E = V_W X_H \tag{3.2}$$

Substitution of Equation 3.2 for P_{XH} in Equation 3.1 and rearrangement using the definition of total SRT in Equation 2.10 gives:

$$\mu_H - b_H = \frac{T_C}{\theta_X T_E} \tag{3.3}$$

Consequently, the net average specific growth rate of the system is inversely proportional to the effective SRT.

$$\mu_H - b_H = \frac{1}{\theta_{XE}} \tag{3.4}$$

The average specific growth rate of heterotrophs, μ_H, which is a function of effective SRT, relates to the average readily biodegradable substrate concentration in the reactor, S_S, with the following rate expression:

$$\mu_H = \mu_{H\max} \frac{S_S}{K_S + S_S} \tag{3.5}$$

The effluent readily biodegradable COD as a growth limiting substrate will possibly be lower than the level that may be calculated using the average specific growth rate in the above equation, depending on the fill time ratio.

The effective SRT exerts a decisive impact on features of practical importance in SBR operation, such as excess sludge production and oxygen requirement as well as the effluent quality. The range of typical SRT values commonly adopted for continuous-flow activated sludge configurations is also applicable for SBR. An important point to remember is that effective and total SRT are different in SBR while they are the same in continuous flow systems. An appropriate effective SRT is often selected based on experience in the case of COD removal from municipal wastewaters. However, for the biological treatment industrial wastewaters and leachates, it is advisable to use kinetic information from treatability studies together with a proper definition of the reactor hydraulics.

3.2.1 Effluent Quality

Process stoichiometry does not allow for the direct calculation of the effluent soluble COD. In domestic sewage treatment for organic carbon removal, selection of an appropriate θ_{XE} does not usually relate to substrate removal, but it is more directed towards producing a flocculent biomass that settles well.

The expression defining the soluble substrate concentration in terms of θ_{XE} has been the major drawback eclipsing the merit of the traditional modelling in design, because it never provided accurate information. In recent modelling approaches however, this expression modified as given in Equation 3.5 yields only the effluent readily biodegradable COD, S_S. For domestic sewage, it is reasonable to assume total removal of the soluble biodegradable COD fractions so that the effluent only contains the total soluble residual COD, S_R incorporating the residual influent inert COD fraction, S_{I1} which by-passes the reactor and soluble inert metabolic products, S_P generated during the course of the metabolic reactions:

$$S_T = S_R = S_{I1} + S_P \tag{3.6}$$

The concentration of the soluble inert products may be calculated in terms of the influent total biodegradable COD and the selected SRT (Orhon andArtan, 1994; Orhon 1998):

$$S_P = f_{SE} b_H \frac{Y_H}{1 + b_H \theta_{XE}} C_{S1} \theta_X \tag{3.7}$$

The stoichiometric coefficient, f_{SE} quantifies the fraction of the endogenous biomass converted into soluble inert products. For domestic sewage, control of the particulate COD in the effluent through efficient settling of the activated sludge is more important for compliance with effluent regulations. The magnitude of f_{SE} and f_{SI} are quite important for high strength industrial wastewaters for the accurate assessment of the effluent fraction that cannot be removed by biochemical processes. A review of experimental results related to soluble inert COD fractions and residual microbial products generation for different industrial wastewaters and domestic sewage is presented elsewhere (Orhon *et al.*, 1999). For industrial wastewaters, the slowly biodegradable COD of soluble nature, S_H should also be considered as a significant component of the effluent soluble COD.

3.2.2. Excess Sludge Production and Reactor Biomass

With the selection of the effective SRT, essential operating parameters can be calculated on the basis of process stoichiometry and relevant wastewater characteristics (i.e. average flowrate and COD fractions). These parameters are basically the same for all activated sludge systems for a selected effective SRT, regardless of the process configuration. The excess sludge production rate, P_{XT} is one of the key parameters and involves a number of particulate components such as active biomass, endogenous residues, influent inert particulate matter of organic and inorganic nature. These components may be integrated into the following expression defining P_{XT} in terms of total suspended solids (g TSS/day), with the assumption that all available biodegradable COD in the influent C_{S1} is depleted:

$$P_{XT} = i_{TSS,COD}\left(Y_{NH} Q C_{S1} + Q X_{I1}\right) + Q X_{FS1} \tag{3.8}$$

In this expression, i_{TSSCOD} is a coefficient to convert COD into TSS; X_{FS1} is the influent fixed solids concentration which can be determined as the difference of influent TSS and VSS values and Y_{NH} is the net heterotrophic yield, accounting both for heterotrophic growth and endogenous respiration, as a function of the effective SRT and can be calculated as:

$$Y_{NH} = (1 + f_E b_H \theta_{XE}) \frac{Y_H}{1 + b_H \theta_{XE}} \tag{3.9}$$

where: Y_H = heterotrophic yield, g COD(g COD)$^{-1}$
f_E = inert fraction of the biomass

The total amount of biomass sustained in the reactor, M_{XT}, is another key parameter in SBR design, which can be computed using the definition of SRT:

$$M_{XT} = V_T X_T = P_{XT}\, \theta_X = P_{XT}\, \theta_{XE} \frac{T_C}{T_C - T_{S+D}} \tag{3.10}$$

The above equation shows that M_{XT} cannot be arbitrarily selected but it is a function of the selected effective SRT and wastewater characteristics. It also indicates that the amount of biomass in the SBR must be higher than that in the aeration tank of a continuous-flow system for the same performance (i.e. same effective SRT). Furthermore, it shows that the process stoichiometry can only determine M_{XT} and not its components V_T and X_T separately, suggesting free selection of one of the variables (volume or biomass concentration) within reasonable limits, based on practical experience.

In continuous-flow systems, the volume of the reaction (aeration) tank is calculated on the basis of a selected mixed liquor suspended solids (MLSS) concentration, because an appropriate value for the MLSS concentration is important not only for reactor sizing, but also for the solids loading which is to be considered in the design of secondary clarifiers. The achievable concentration in the recycle flow, X_R, is determined as a function of settling characteristics of sludge and solids loading on the settling tank. With the estimation of an appropriate X_R value, the recycle ratio, R, required to maintain the desired MLSS concentration in the reactor, X_T, is calculated from the mass balance for solids around the settling tank

$$R = \frac{X_T}{X_R - X_T} \tag{3.11}$$

The recommended range for the MLSS concentration (2500 – 5000 g/m^3) yields sludge recycle ratios of less than 1.0 in common operating practice. Consequently, the interaction between the bioreactor and the secondary clarifier is accounted for through selection of an appropriate MLSS concentration. On the other hand, in SBR design, this relationship should be analyzed in a different way as outlined below, since biological conversions and settling occur in the same tank.

3.3. SINGLE TANK DESIGN PRINCIPLES

In SBR design sizing the reactor is usually based on the selection of T_C, and hence the hydraulic retention time, HRT, instead of selecting a suitable MLSS concentration. The choice of an appropriate T_C value must be justified by means of the achievable settling characteristics of activated sludge expected to occur during the settle phase, as well as mixing and oxygen transfer considerations. The first constraint is the initial volume, V_0, which should be large enough to hold the settled biomass with a safety factor, SF:

$$V_0 = SF\, V_S = SF\, \frac{M_{XT}}{X_R} \tag{3.12}$$

It is obvious that smaller Vs and hence V_0 levels may be obtained for higher settled sludge concentrations, X_R. The maximum achievable value of X_R is the function of settling characteristics of sludge and the duration of settling phase, T_S; it can be estimated from a selected value for the sludge volume index, SVI (ml/g), defining the volume (in ml) occupied by a gram of sludge after 30 minutes of quiescent settling. From this definition,

$$X_R\,(g/m^3) = \frac{10^6}{SVI} \tag{3.13}$$

and using the value of M_{XT} in Equation 3.10, V_0 may be re-expressed as:

$$V_0 = SF\, P_{XT}\, \theta_X\, SVI\, 10^{-6} \tag{3.14}$$

Introducing V_F through basic definitions of Q, the recycle ratio for SBR systems, (V_0/V_F), can be calculated directly as a function of m (i.e. number of cycle per day):

$$\frac{V_0}{V_F} = SF\,(P_{XT}/Q)\,\theta_X\, SVI\,10^{-6}\, m \tag{3.15}$$

or using the definitions of effective SRT and TC:

$$\frac{V_0}{V_F} = SF\,(P_{XT}/Q)\, SVI\,10^{-6}\, \frac{\theta_{XE}}{(T_E/T_C)}\, \frac{24}{T_C} \tag{3.16}$$

The above expression indicates that the selection of T_C and its effective fraction determine the recycle ratio, V_0/V_F, and in turn, calculation of HRT from Equation 2.3, for a given set of wastewater and sludge characteristics and effective SRT. Fill volume requirement increases with increasing T_C; however, higher T_C does not always mean higher HRT. Because, V_0/V_F is inversely proportional with T_C; decreasing T_C also results in decreasing the effective fraction, T_E/T_C, since the time to be devoted to settling and draw is almost constant for all selected cycle times. Therefore, HRT may increase with decreasing T_C, due to a higher V_0/V_F value as a result of lower T_C and T_E/T_C fraction as can be seen from Equation 3.16. Thus, for each set of operation condition dictated by wastewater strength and effective SRT, a cycle time, T_C that will minimize the total volume requirements can be determined.

With the simplifying assumption that the amount of biomass at the end of the react phase, remains the same after solids settling, the MLSS concentration in the reactor after fill, X_T, can be calculated as a function of recycle ratio:

$$X_T = \frac{V_0/V_F}{(1+V_0/V_0)}\,\frac{X_R}{SF} \tag{3.17}$$

It is well known that SBR can be operated with high MLSS concentrations essentially due to high sludge recycle ratios which can be obtained without any cost since they simply corresponds to V_0/V_F. Optimum T_C values usually yields recycle ratios of higher than 1.0 depending on wastewater strength and effective SRT used. In the case of high strength wastewater treatment, T_C values that will yield high recycle ratios up to an acceptable limit (for example, $V_0/V_F < 4.0 - 5.0$) should be selected to minimize HRT. For very high recycle ratios, resulting high MLSS levels may require higher settling time since the initial settling velocity is a function of sludge concentration which will, in turn, require longer cycle times.

An example of the relationship between selected T_C and HRT is shown in Table 3.1, calculated for three influent total COD levels of 300 mg/l, 600 mg/l and 900 mg/l for an SBR operation at two different effective SRT values of 10 d and 20 d, using relevant data on wastewater characteristics and operating parameters commonly adopted for domestic sewage as given in Table 3.2.

Table 3.1 shows that V_0/V_F is quite sensitive to T_C while the resulting HRT remains approximately constant for a range of increasing T_C values, indicating an optimum T_C range of 4 to 6 h for a weak domestic wastewater. In this range, HRTs are relatively constant whereas recycle ratios and effective fractions vary significantly. This reflects an additional flexibility in SBR design, giving the possibility of selecting the most appropriate T_C value within the optimum range without appreciably affecting HRT, as required for the expected performance of the process. This aspect will be further elaborated in the next chapter on nutrient removal. It is also noteworthy that the optimum cycle time values increase with increasing θ_{XE} and wastewater strength.

Table 3.1 Relationship between selected cycle time, recycle ratio and the hydraulic retention time.

(a) *effective SRT=10d,*

	C_{TI}=300 mgCOD/l		C_{TI}=600 mgCOD/l		C_{TI}=900 mgCOD/l	
T_C h	V_0/V_F -	HRT h	V_0/V_F -	HRT h	V_0/V_F -	HRT h
4	1.53	10.1	3.06	16.2	4.59	22.4
5	1.02	10.1	2.04	15.2	3.06	20.5
6	0.77	10.6	1.54	15.2	2.31	19.8
7	0.61	11.3	1.22	15.5	1.83	20.3
8	0.51	12.1	1.02	16.2	1.53	20.4
9	0.45	13.1	0.90	17.1	1.35	20.7
10	0.38	13.8	0.77	17.7	1.14	21.4
11	0.34	14.7	0.68	18.5	1.02	22.2
12	0.31	15.7	0.62	19.4	0.93	23.2

(b) *effective SRT=20d,*

	C_{TI}=300 mgCOD/l		C_{TI}=600 mgCOD/l		C_{TI}=900 mgCOD/l	
T_C h	V_0/V_F -	HRT h	V_0/V_F -	HRT h	V_0/V_F -	HRT h
4	2.52	14.1	5.04	24.2	7.56	34.2
5	1.68	13.4	3.36	21.8	5.04	30.2
6	1.26	13.6	2.52	21.1	3.78	29.4
7	1.01	14.1	2.02	21.1	3.03	28.2
8	0.84	14.7	1.68	21.4	2.52	28.2
9	0.72	15.5	1.44	22.0	2.16	28.4
10	0.63	16.3	1.26	22.6	1.89	28.9
11	0.56	17.2	1.12	23.3	1.68	29.5
12	0.50	18.0	1.00	24.0	1.50	30.0

Table 3.2 Selected wastewater characteristics and operating parameters

Parameter	Symbol	Value	Unit
Heterotrophic yield coefficient	Y_H	0.64	gCOD(gcell COD)$^{-1}$
Heterotrophic endogenous decay coef.	b_H	0.15	day^{-1}
Particulate inert fraction of biomass	f_E	0.20	-
TSS to cell COD	$i_{TSS,COD}$	0.9	gTSS/gCOD
Biodegradable fraction of COD	f_{CS}	0.80	-
Particulate inert COD fraction	f_{XI}	0.10	-
Fixed Solids/Total COD of influent	f_{FS}	0	gTSS/gCOD
Sludge volume index	SVI	100	ml/gTSS
Duration of settle, draw and idle phases	T_{S+D+I}	2	hr
Safety factor	SF	1.3	-

Selection of T_C and hence the reactor volume, also require consideration of the constraints related to the rate of oxygen transfer rate, as outlined in the following section.

3.4. AERATION SYSTEM DESIGN PRINCIPLES

As previously discussed, the effective SRT also determines oxygen requirement at steady state for a given wastewater. The heterotrophic oxygen requirement, OR_H can be calculated as:

$$OR_H = (1 - Y_{NH})QC_{SI} = mV_F(1 - Y_{NH})f_{CS}C_{TI} \tag{3.18}$$

The OR_H value calculated from the expression above should be considered for the most critical wastewater temperature. This requirement must be satisfied with the fraction of the oxygen supply that is available to the heterotrophic biomass. As the oxygen transfer efficiency, SOTE, is commonly defined under standard conditions, OR_H should first be converted into standard oxygen requirement, SOR. The required air supply to the reactor may then be calculated as a function of SOR and SOTE. Details of this evaluation are beyond the scope of this report and can be found elsewhere (Grady et al. 1999; Orhon and Artan, 1994).

The specific characteristics of SBR operation dictate that the daily amount of oxygen demand must be supplied to the reactor during aerated periods of the cycle. Accordingly, an oxygen requirement rate, AOR, must be defined for the effective portions of the daily operation:

$$AOR = OR_H / mT_A = \frac{OR_H}{24(T_A / T_C)} \tag{3.19}$$

For a given type of aeration device, the total power input required for oxygen transfer will be determined by this quantity of oxygen which is essentially independent of the reactor volume. Generally, for economic reasons, the same aeration equipment is used to satisfy the mixing requirements necessary to keep the MLSS in suspension. A minimum volumetric power input is required for mixing. Thus, the volumetric oxygen transfer rate at the maximum water level, r_{OT} ($gO_2/m^3.h$) must be considered in design. This parameter can be estimated from the following basic mass balance, neglecting the difference between influent and effluent DO concentrations:

$$V_T r_{OT} = OR_H / mT_A \tag{3.20}$$

The volumetric oxygen transfer rate, r_{OT} may be redefined in terms of C_{S1} and HRT, combining equating Equations 3.20 and 3.18:

$$r_{OT} = \frac{(1-Y_{NH})QC_{S1}}{V_T mT_A} = \frac{(1-Y_{NH})f_{CS}\,C_{T1}}{HRT(T_A/T_C)} \tag{3.21}$$

Constraints associated with r_{OT}, should also be observed for the selection of T_C, and hence the reactor volume, in addition to considerations outlined in the previous section. It is evident from the above equation that higher r_{OT} levels are required for shorter T_C values due to shorter aerated time fractions (i.e. T_A/T_C) for the same HRT. This may somewhat offset the benefit in the total reactor volume requirement acquired by decreasing T_C. The advantage of operating with a smaller reactor volume will have to be evaluated and compared, especially for weaker wastewaters, with the additional expense in aeration equipment due to longer fractional time devoted to settling and draw and consequently less time available for aeration.

Volumetric power input should satisfy two conditions: it must be (i) lower than a critical upper value to avoid disruption of biological flocs and deterioration of settling properties, and (ii) high enough to sustain the necessary mixing conditions in the reactor. Suggested values for the upper limit range from 40 - 80 gO_2/m^3-hr depending on the type of aeration equipment. This limit must be observed for the initial reactor volume (i.e.V_0) if aeration starts together with filling as usually practised for carbon removal systems. Rewriting Eq 3.21 for V_0, the volumetric oxygen transfer rate at the minimum water level, $r_{OT@0}$, will be:

$$r_{OT@0} = \frac{(1-Y_{NH})mV_F C_{S1}}{V_0 mT_A} = \frac{(1-Y_{NH})f_{CS}C_{T1}}{T_A\,(V_0/V_F)} \tag{3.22}$$

Inserting the value of V_0/V_F in Equation 3.16, $r_{OT@0}$ may be redefined as a function of the effective SRT, independently from the selected T_C:

$$r_{OT@0} = \frac{(1-Y_{NH})f_{CS}}{SF(P_{XT}/Q)\theta_{XE}\,SVI10^{-6}} \tag{3.23}$$

Recommended air input to satisfy mixing requirements for diffused air systems is in the range of 20–30 m^3 air/min per 1000 m^3 of reactor volume, and 15-30 W/m^3 of reactor volume for vertical shaft mechanical aerators; This corresponds to 15-30 gO_2/m^3-hr with the assumption that mechanical aerators transfer approximately 1 gO_2/W-hr to the mixed liquor in activated sludge systems. If the calculated r_{OT} is lower than this value, the mixing requirements will control the selection of the aeration equipment, a situation often encountered in systems operated at high SRT values. In SBR systems, r_{OT} can also be increased by selecting a lower T_C if required to obtain a lower HRT. Consequently, T_C should be selected so as to optimize the requirements for both reactor volume and aeration capacity until mixing requirements control the sizing of the aeration equipment. Table 3.3 reflects the relationship between the selected T_C and the volumetric oxygen transfer rates at both maximum and minimum water levels, for the same wastewater characteristics and relevant operating parameters given in Table 3.1.

On the basis of general principles of gas transfer, r_{OT} may be defined in terms of the oxygen transfer coefficient, K_La (T^{-1}):

$$r_{OT} = K_L a(S_{OS} - S_O) \tag{3.24}$$

where, S_{OS} defines the saturation concentration of dissolved oxygen at process conditions (gO_2/m^3), and S_O, the actual dissolved oxygen concentration in the reactor during aerated periods (gO_2/m^3). With the initial assumption of a constant DO concentration during the aeration period, a K_La can be estimated from Equations 3.24 and 3.21. In actual operation however, the oxygen uptake rate, OUR will vary as dictated by the rate of biochemical reactions during the aerated process phase and a constant K_La, will yield a DO profile according to the mass balance below:

$$\frac{dS_O}{dt} = K_L a(S_{OS} - S_O) - OUR \tag{3.25}$$

Table 3.3 Effect of selected T_C on volumetric oxygen transfer rates at min. and max. volumes

(a) *effective SRT=10d,*

T_C	C_{TI}=300 mgCOD/l		C_{TI}=600 mgCOD/l		C_{TI}=900 mgCOD/l	
	$r_{OT@0}$	r_{OT}	$r_{OT@0}$	r_{OT}	$r_{OT@0}$	r_{OT}
h	$gO_2/m^3.h$	$gO_2/m^3.h$	$gO_2/m^3.h$	gO_2/m3.h	$gO_2/m^3.h$	$gO_2/m^3.h$
4	52.4	31.9	52.4	39.8	52.4	43.1
5	52.4	26.6	52.4	35.3	52.4	39.2
6	52.4	22.8	52.4	31.8	52.4	36.6
7	52.4	19.9	52.4	29.1	52.4	33.3
8	52.4	17.7	52.4	26.5	52.4	31.5
9	52.4	15.8	52.4	24.2	52.4	30.0
10	52.4	14.6	52.4	22.7	52.4	28.2
11	52.4	13.4	52.4	21.3	52.4	26.6
12	52.4	12.3	52.4	19.9	52.4	24.9

(b) *effective SRT=20d,*

T_C	C_{TI}=300 mgCOD/l		C_{TI}=600 mgCOD/l		C_{TI}=900 mgCOD/l	
	$r_{OT@0}$	r_{OT}	$r_{OT@0}$	r_{OT}	$r_{OT@0}$	r_{OT}
h	$gO_2/m^3.h$	$gO_2/m^3.h$	$gO_2/m^3.h$	gO_2/m3.h	$gO_2/m^3.h$	$gO_2/m^3.h$
4	35.6	25.5	35.6	29.7	35.6	31.6
5	35.6	22.4	35.6	27.5	35.6	29.8
6	35.6	19.9	35.6	25.6	35.6	27.6
7	35.6	17.9	35.6	23.9	35.6	26.8
8	35.6	16.3	35.6	22.4	35.6	25.5
9	35.6	14.9	35.6	21.0	35.6	24.4
10	35.6	13.8	35.6	19.9	35.6	23.4
11	35.6	12.8	35.6	18.9	35.6	22.4
12	35.6	12.0	35.6	18.0	35.6	21.6

In this mass balance equation, OUR should be expressed as below, as a function of the heterotrophic growth and endogenous respiration:

$$OUR = \left(\frac{1-Y_H}{Y_H}\right)\mu_{Hmax}\frac{S_S}{K_S+S_S}X_H + (1-f_E)b_H X_H \tag{3.26}$$

The effect of K_La on the DO profile is illustrated in the following exercise, involving an influent total COD of 600 mg/l with a biodegradable fraction of 80% as in the previous examples, and an SBR operation with an effective SRT of 10days, a cycle time, T_C of 6 hr and two different FTR values of 0.5 and 1/6. The resulting DO profiles are plotted in Figure 3.1 for three different K_La values: A K_La value of 110 d^{-1} was first estimated from Equation 3.24 assuming a constant DO level of 2.0 mg/l and using the calculated r_{OT} value from Table 3.3 for selected operating conditions. Then, two K_La values of 80 and 160 d^{-1} were additionally selected for the evaluation. Fig 3.1 (a) shows that the estimated K_La (110 d^{-1}), induces DO concentrations below 2 mg/l during the aerated period with filling, which then increases up to 3.5 mg/l until aeration ceases. On the other hand, a lower K_La value of 80 d^{-1} results in DO concentrations below 2 mg/l during entire aerated period. Furthermore, as discussed in the section 2.3, the fill time ratio, FTR, exerts a significant effect on substrate concentration and hence on the OUR profile. Therefore, a lower FTR causes a much larger variation in the DO concentration profile during the aeration period, as shown in Figure 3.1 (b).

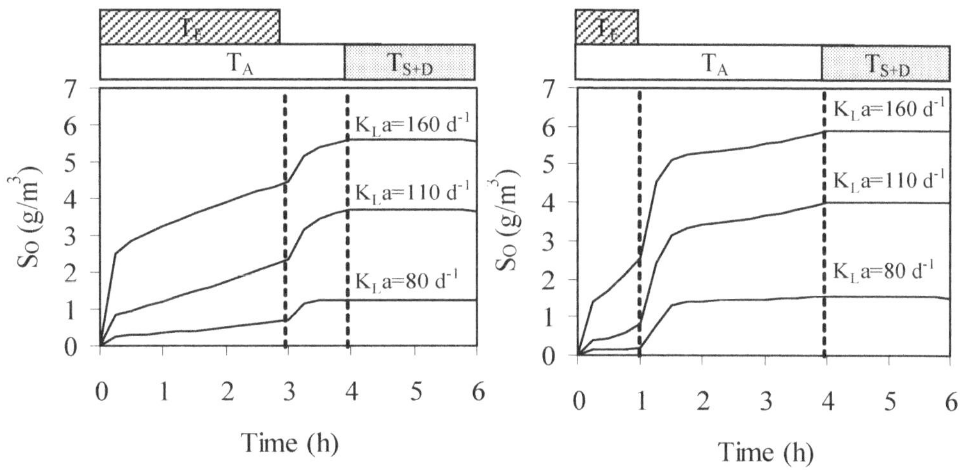

Figure 3.1. Effect of K_La on DO profiles at a) FTR=0.5, b) FTR=1/6

3.5. PROCESS DESIGN

3.5.1. Design Procedure

A step-wise design procedure for COD removal in SBR systems, based on the selection of an appropriate effective SRT, is given in Figure 3.2. Each step of this iterative procedure is briefy explained below:

1. *Collect the basic design data*:

- Wastewater flow rate, with the daily average, Q, and peak values, daily variations and seasonal or specific fluctuations.
- Daily average concentrations of influent COD, C_{T1} and its fractions; mainly the total biodegradable COD, C_{S1}, the particulate and soluble inert COD, X_{I1} and S_{I1}. For domestic sewage, if specific information on COD fractionation is not available,

$C_{S1}/C_{T1} = 0.85$; $S_{I1}/C_{T1} = 0.05$ and $X_{I1}/C_{T1} = 0.10$ may be adopted as commonly accepted values in the literature. A specific COD fractionation is essential for industrial wastewaters.

- Other wastewater characteristics, such as TKN, TP, TSS, VSS, pH, alkalinity, temperature and inhibitors if necessary.
- Kinetic and stoichiometric coefficients: COD removal from municipal wastewater does not usually require extensive kinetic information. For essential stoichiometric parameters like Y_H and f_E, the endogenous respiration rate coefficient, b_H and conversion factors, representative values can be assumed. However, determination of wastewater–specific coefficients is required in the case of industrial effluents and special wastewaters.
- Applicable efluent discharge limitations.

2. *Select effective SRT*- θ_{XE} should be selected so as to satisfy the effluent requirements. For domestic sewage, selection is generally based on practical experience to sustain a biomass with good settling properties: A θ_{XE} range of 4–8 days is typical for warm weather conditions; it may be increased up to 15 days for cold weather operation. Effluent soluble biodegradable COD is generally of no concern since it has a minor contribution to the total effluent COD. Many factors, such as sludge production, oxygen requirement and settling problems, play more significant roles in its selection.

3. *Estimate effluent COD*- Effluent soluble biodegradable COD, S_S and S_H can be estimated if kinetic information is available. Soluble inert products, S_P, and/or soluble inert COD of influent origin, S_I, will leave the system and directly contribute to the effluent soluble COD. Escaping biomass in the effluent stream will contribute to the total COD of the effluent.

4. *Calculate P_{XT}* – The daily amount of sludge produced is calculated from Equations 3.8 and 3.9.

5. *Calculate OR_H* – The daily amount of oxygen requirement is calculated from Equation 3.18.

6. *Select T_C* – T_C should be selected so as to optimize the requirements for reactor volume and aeration equipment. Iteration should preferably be started from an acceptably low value (for example 4 h).

7. *Select T_S, T_D and T_I* – A cycle has to include an additional time after process phase for settling and draw. Durations range from 0.5 to 2 hours for each of these phases, depending on factors like the employed method of decantation, sludge settling velocity etc. The idle phase acts as an equalization tank in the reactor to control the system during variable flow conditions; it can be omitted in SBR systems that have a holding tank.

8. *Select SVI; calculate V_0/V_F* – Settled sludge concentration or volume, (SVI) and the safety factor must be selected to calculate V_0/V_F from Equation 3.10. A selected T_C value corresponding to a V_0/V_F range of 1.0 – 4.0 will probably minimize HRT depending on wastewater biodegradable COD content and the selected SRT. Any V_0/V_F in this range can be acceptable unless the circumstances impose an overriding design priority (this issue will be discussed in detail in the following nutrient removal chapter).

9. *Calculate HRT and V_T* – HRT is calculated from Equation 2.3 also enabling the calculation of the total volume, V_T. An iterative evaluation must be made to ensure that the calculated HRT represents the lowest achievable level.

10. *Calculate the volumetric oxygen transfer rate*- r_{OT} is calculated from Equation 3.15 and checked with respect to mixing requirements; the calculated r_{OT} value must be high enough to sustain the necessary mixing conditions in the reactor. If not, it can be adjusted by selecting a new T_C. The ultimate decision for the selection of an appropriate T_C must be justified with cost estimations for the rector and the aeration equipment.

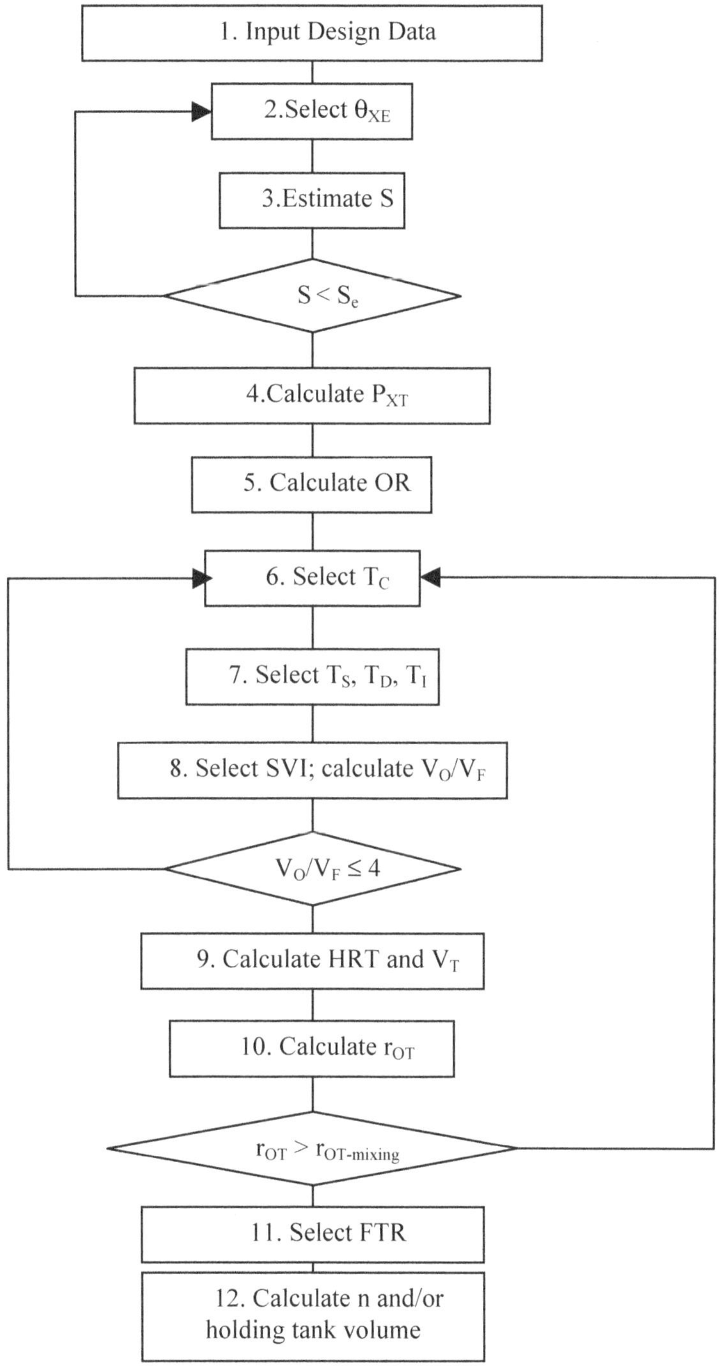

Figure 3.2.The design algorithm for carbon removal.

11. Select FTR- FTR is selected with due consideration of reaction kinetics and population dynamics, as discussed in the previous chapter.
12. Calculate n and/or holding tank volume-Number of reactors is selected in accordance with the selected FTR and the volume of each parallel reactor is calculated as V_T/n. Reactor dimensions (surface area, height of the clear water zone, etc.,) are determined for an appropriate reactor height. If necessary,, the holding tank is dimensioned on the basis of the daily wastewater flow hydrograph and the selected filling pattern.

3.5.2. SBR Equipment and the Practical Aspects of SBR Design

3.5.2.1. Factors Affecting Reactor and Equalization Tank Dimensioning

In professional practice, designing an SBR system is, in essence, minimizing the total investment and operating costs, without compromising the targeted level of process efficiency and reliability. The critical step in design is selecting a T_C, which in turn will enable calculating the total reactor volume and oxygenation capacity as discussed in the previous sections. T_C should be optimally selected such that the total construction, equipment, piping, cabling, instrumentation, control, etc and the operating costs are minimized.

The total reactor volume can be divided into as many reactors as needed to minimize the total investment and operational costs of the total tankage and related equipment of the reactors and an influent equalization/pumping tank. Where hydrograph and the number of reactors allow, merely an influent pump pit, with minimal storage capacity can be used. Most of the time, though, some equalization will prove profitable, duly equipped with mixing/aeration devices to prevent adverse conditions during wastewater storage. In larger plants legal requirements may impose choosing the number of reactors such that a redundant reactor is designed into the system, enabling regular or emergency servicing of one of the reactors, while the remaining reactors handle the waste flow with no significant impairment of the effluent quality. In smaller plants, the widespread practice is employing one reactor with some equalization, or two reactors with practically no equalization.

Marked diurnal, weekly or seasonal fluctuations should also be taken into account in designing SBRs. Generation of wastewater in only one or two shifts per day is quite frequent with small industrial plants, as with the pronounced variation of sewage flow from small communities or holiday resorts. An SBR operating one cycle per day, settling and decanting during the low flow period usually furnishes a practical solution. There is also an appreciably widespread use of compact package plants, designed especially for small communities, mostly practising continuous fill and several discharges per day. With large seasonal variation, it is customary to design multiple reactors, some operating seasonally to keep pace with wastewater generation.

Where SBR needs to be in succession with tertiary treatment systems like tertiary chemical treatment or granular media filtration, measures generally have to be taken to maintain some uniformity of flow to tertiary treatment steps. Such requirement will act as an additional factor in the determination of number of reactors in an SBR plant. The optimal solution usually involves employing some equalization after the SBRs, probably in addition to an adjustment of the number of reactors.

Another important factor in determining the number of reactors and the equalization volume is the selection of the fill time ratio (FTR). Especially where dump fill is required for substrate accumulation or selector purposes, it should be noted that the equalization requirements will increase drastically with decreasing number of reactors. Where FTR is known to have no significant effect on sludge characteristics, priority could be given to minimizing total tankage costs including the equalization tank. In this case FTR will be determined largely by the number of reactors and by equalization requirements.

Another significant issue associated with FTR is its effect on the distribution of oxygen demand throughout the aerated period, with demand peaking in the early stages of the period in systems operating with low FTRs. Particularly when fostering growth of floc-forming species is aimed with a front-loaded, low FTR operating mode, the oxygenation equipment must be sized accordingly to prevent low dissolved oxygen concentrations during peak demand.

After determination of total reactor volume and the number of reactors, the reactor geometry remains to be selected. In principle, cylindrical tanks are cheaper to build, but multiple reactors are cheaper to build in prismatic form, thanks to possibility of building them with adjacent walls. Cylindrical tanks may require less installed mixing power to keep mixed liquor in suspension, but may need baffles to prevent mass rotation hindering homogenization. Furthermore, multiple units of aeration equipment are more conveniently and efficiently configured in prismatic tanks. In larger plants the SBR reactors are almost invariably prismatic tanks. Sometimes the type of SBR operation dictates the reactor geometry. In SBRs with incessant fill, the tanks are usually built as rectangular prisms, to minimize short-circuiting of the influent during decantation, notwithstanding the baffle configuration built in such applications.

Mixed liquor depth is another issue to be addressed in the design phase. This depends on a multitude of factors including land cost, construction costs, comparative prices of various mixing/aeration equipment, energy price, etc. Generally the most cost effective solution is building deep reactors with typically 6-9 m maximum operating depth that occupy less land and allow using state-of-the-art aeration/mixing equipment under their higher efficiency operating range, as the bottom aeration systems deliver higher efficiencies at greater depths. Usually such advantages are of greater significance in larger plants. As plants get smaller, especially in cases where land use is not a critical issue, decreasing the level of sophistication and simplifying operation and maintenance may override the efficiency concerns. In such cases shallow tanks are built, often as concrete or membrane-lined earthen lagoons with trapezoidal section, typically equipped with floating surface aerators. Surface aerators and self-aspirating aerators have a maximum operating water depth of 3-7 m depending on equipment size, whereas diffusers and submerged aerator/mixers, for instance, can operate at approximately twice this range.

Regarding tank depth, a crucial issue worth mentioning is the overwhelming advantage of the SBR over continuous-flow systems in cramped places where floor area is at a premium. Whereas the reactors can be built to any height, thus minimizing the land requirement, the continuous-flow sedimentation tanks have no such elasticity, rendering continuous-flow systems unaffordable under such circumstances or even impossible where land restriction is absolute.

The reactor material is reinforced concrete or coated mild steel most of the time, though other materials like stainless steel, glass reinforced polyester (GRP), etc are also used. Where lagoons serve as reactors, they are generally geo-membrane or concrete lined earthen structures. In cold climates reactors may have to be built with covered tops.

3.5.2.2. Mixing/Aeration Equipment Used in SBRs

A variety of mixing/aeration equipment is used in SBR plants. Bottom mounted systems comprising fine-bubble diffuser systems and low speed submerged turbines constitute a distinct high efficiency cluster as compared to various other aeration systems.

Diffused air aeration systems once having the drawback of frequent clogging in intermittent operation characteristic of SBRs, have long since developed non-clog elastomeric membrane type diffusers suitable for discontinuous operation. Still, faster deterioration of efficiency and more downtime should be expected when using diffuser systems, as compared to many mechanical systems. Blowers supply air to the system, and propellers are to be installed in the reactor to maintain suspension and homogenization of the mixed liquor where unaerated mixing is to be practiced. In addition to an increase in oxygen transfer efficiency by increasing the water depth, the efficiency also increases due to greater coverage of the reactor bottom area by the same number of diffusers in a deeper reactor with the same volume. Consequently, deeper reactors with typically 8-10 m maximum operating water depth make better use of diffused air aeration systems. Diffuser systems generally call for evacuation of the reactor for major servicing and maintenance; therefore provision should be designed into the system to avoid discharge of the mixed liquor.

In contrast to diffuser systems, some of the submerged aerator/mixer systems are endowed with the hoist-in-place feature rendering their maintenance extremely practical. Operating up to 15 m water depth, they consist of a low-speed submersible mixer of special design, placed on the reactor bottom. They are capable of mixing the reactor contents during unaerated periods, and during the aerated periods air fed into the rotor by the blowers is dispersed into small sized bubbles and the mixed liquor/air dispersion is ejected from the unit horizontally/radially, effecting high efficiency oxygen transfer.

The so called hyperboloid mixer systems comprising large, nearly conical GRP impellers suspended by long shafts have their drive units mounted on fixed spanning bridges over the reactor rendering their maintenance very practical. Similar to the submerged aerator/mixers, they are capable of unaerated mixing, and during aerated periods, air fed beneath the mixer is dispersed into small sized bubbles by the shear forces of the low-speed, high-flow mixing action. Often, dual-speed motors are used to tackle the increased mixing power requirement during aeration. Hyperboloid mixers tend to decrease the surface scum by sucking it into the vortex created along the mixer shaft and dispersing it into the bulk mixed liquor.

Surface or bottom-mounted self-aspirating aerator/mixers consist in a high speed propeller which creates elevated mixed liquor velocities at the propeller tip. Taking advantage of high fluid velocities and the consequent low pressure, air is drawn into the mixed liquor through the hollow propeller shaft or some other suction pipe, and dispersed into small sized bubbles by the high shear forces at the propeller tip. Bottom-mounted types operate up to some 7 m water depth. Airflow is adjusted by throttling the air intake. Mixing and aeration efficiencies of aspirating aerators are not as well documented as the other types of aeration equipment. An additional drawback is proneness to floc disruption due to high impeller speeds.

Jet aeration systems are finding widening use especially in smaller SBR sytems. The system comprises motive pumps, mostly installed outside the reactor, recirculating the mixed liquor through a manifold pipe equipped with venturi nozzles to provide complete mix conditions in the reactor. During aerated periods, air blowers dose air to the nozzles to maintain desired level of oxygen concentration in the reactor. Jet aerators possess the remarkable advantage of having no submerged moving parts. On the other hand, their oxygenation efficiencies are somewhat less than diffused air aeration systems or submerged aerator/mixers, and dependent to a great extent to the air/liquid ratio. To prevent clogging of nozzles, jet aerators normally require fine screening of the wastewater where primary settling is not practiced.

Variation of water depth must be taken into account when designing bottom-mounted aeration systems. Effects of drifting transfer efficiency during filling should be calculated, and care should be taken to avoid exceeding maximum allowable power input or maximum possible oxygen transfer rates per unit volume of mixed liquor, particularly around minimum water depths. Such caution becomes even more critical where SBRs are intended to be operated as aerobic selectors during a part of their process cycles, exercising dump fill. Furthermore it should be noted that SBRs operated as aerobic selectors will require much larger installed aeration capacity as compared to equivalent continuous-flow systems whereas the power consumption is, in principle, equal.

Floating, vertical-shaft, low-speed, mechanical aerators offer a viable aeration alternative for carbon removal and nitrification systems. Though not comparable to most bottom-mounted systems in oxygen transfer efficiency, they find widespread use especially in smaller plants, thanks to their simplicity, and rugged nature. In spite of lower cost, direct drive high speed surface aerators are not as desirable due to lower efficiency and their tendency to disrupt the floc structure of the activated sludge. Although surface aerators are intrinsically unfit for unaerated mixing, there are many nutrient removal SBR systems available on the market equipped with surface aerators which operate by intermittently switching on the aerator for short durations to homogenize the tank contents during the unaerated mix periods.

The specifications and guidelines given by surface aerator manufacturers concerning geometry of aeration tanks equally apply to SBR reactors. Compliance with the minimum/maximum depth and/or side/depth ratio range specifications is more critical for an inherently variable-depth system like the SBR, particularly as the depth variation becomes larger. With V_0/V_F ratios around 1 and lower, it may become problematic to satisfy some manufacturers' specifications at all operating depths, and resort to special measures like adding a draft cone to the aerator or building flow-directing structures at the reactor bottom might become mandatory. Reactors with trapezoidal cross-section alleviate this problem and provide some relief as they have a more favourable volume to depth ratio, compared to prismatic tanks. Also manufacturers' specifications and guidelines regarding effective mixing radius, variation of SOTR with tank volume and configuration of multi-aerator tanks should be carefully taken into account in reactor and process design.

3.5.2.3. Other SBR Equipment

All types of equipment and instrumentation used in the SBRs are also used in continuous-flow activated sludge systems with the exception of decanters, not accounting for their limited use in some batch-operated sludge holding/thickening tanks. Decanters have a central function in the SBR process: they draw and discharge, during the draw phase, the clear supernatant overlying the settled sludge, bringing the reactor volume back to Vo.

Although simpler schemes such as decant ports placed at an appropriate height on the reactor wall, or stationary decant/discharge pumps are used to some extent in smaller plants, moving decanters enjoy the most widespread use in the SBR practice. In contrast to stationary, fixed point decanting devices which can only commence supernatant discharge sometime after the sludge blanket descends past the discharge point, i.e. after the settling is virtually completed, moving decanters can start much sooner, moving with the descending water level and drawing supernatant from the surface. The decanter head assembly consisting of one or more ports to withdraw water, and a baffle structure to prevent access of surface scum to the ports is positioned at the water surface either by integral floats (i.e. a floating decanter), or by an electric drive lowering the assembly at a constant or pre-determined speed. The decanter head assembly is connected via a flexible hose, or rigid piping and swivel joints, to a flanged discharge hole in the reactor wall. Where gravity discharge is not possible, a discharge pump is added to the system, either installed outside the reactor or incorporated into the head assembly. When decanting is over, the decantation head assembly is generally winched out of the water to avoid entrance of sludge into the decanter until the next draw phase. Decanter ports, pipes, available head or discharge pumps need to be appropriately selected and sized in order to draw the maximum V_F within the draw phase without disturbing the sludge layer. Particularly with low V_o/V_F ratio and deep reactor systems, settle and draw times and decant rate should be carefully adjusted to avoid suction of sludge.

Some SBR plants are equipped with skimming systems, mostly similar to decanters, for abatement of nuisance due to surface scum and foam. Experience has shown that separate treatment, for instance in drying beds, and disposal of collected scum gives better results than returning it to the headworks. Wasting of surplus sludge is carried out by gravity or by pumps installed in or outside the reactor, generally several times a week during a draw phase.

3.5.2.4. Control of SBR Systems

Since the SBR is a batch process, all the events take place repeatedly in an on-off manner: fill, mixing, aeration, draw and sludge wasting. This has made it preferable, to say the least, from the beginning, to employ some sort of automation. Starting historically with hard-wired control systems comprising level-switch and timer-relays and interlocks, SBR systems have been controlled with programmable logic controllers (PLC), and especially in case of larger plants they are now equipped with computer-based control systems, with the supervisory control and data acquisition system (SCADA) feature which even makes it possible to remotely monitor and intervene in the process. The equipment, instrumentation and software frequently involved in carrying out the basic process steps and functions are briefed below:

Fill: Control valve starts and stops fill on a time-controlled basis; in case of high flow, data from reactor level sensor may stop fill in advance overriding the time-control, and may also trigger shifting to high-flow operational mode, modifying cycle/period times in all reactors, etc. If post-denitrification by intermittent fill is practised, fill control valve may be controlled by data from dissolved oxygen (DO) and oxidation-reduction potential (ORP) sensors in addition to time and level.

Mixing/aeration: Processing data from timers, level, DO and ORP sensors, mixing and aeration are controlled in an on-off or proportional control fashion. Continuous or discrete proportional control can be effected through changing the number of operating units (e.g. blowers), shifting speed in dual-speed drives (e.g. blowers, hyperboloid mixers), throttling air intake of self-aspirating aerators, turning up or down frequency inverters (e.g. blowers, vertical shaft aerators), etc.

Draw: Draw is started mostly on a time-control basis by opening the related control valve, starting a decant pump, winching down a floating decanter, etc. End of decant is signalled by a level sensor set at V_0 generally with an overriding timer-control and alarm function. A turbidity/SS sensor on the discharge line may alarm suction of sludge. Such sensor moving with, but placed at a safe distance below the decanter head may act as a safeguard against decanting too close to the sludge/supernatant interface.

In larger plants, the control system may continuously monitor the process status to display numerically/graphically phase/period, time remaining to end of phase/period, reactor mixed liquor level and volume, DO, ORP, valve positions, equipment status, etc. The control software should include contingency programs for scenarios like wet weather flow (WWF), flow in excess of WWF, valve failures, process upsets, etc. Especially for complicated SBR plants like large nutrient-removal facilities, a sophisticated control-automation system is probably a greater necessity than an equivalent continuous-flow system.

3.5.3. Design Example

This section presents an illustrative example for the design of carbon removal SBR systems. Some aspects of SBR design like flow equalization and practical constraints associated with aeration equipment will be dealt with here and not repeated in similar design examples for nutrient removal systems. The design involves sizing of SBR reactor(s) and equalization tank for an industrial wastewater with (a) diurnally uniform flow; (b) uniform flow during one shift (8 h) per day with no flow the rest of the day, and calculation of the necessary low speed surface aerator power, assuming an AOR of 1.3 kg O_2/kWh. The design will follow the stepwise procedure given in Figure 3.2.

1. *Available design data.* - The following data are available for an industrial wastewater with 1000 m3/day design flow:

C_{T1}=1900 mgCOD/l	X_{I1}= 200 mgCOD/l
C_{S1}=1600 mgCOD/l	S_{I1} = 100 mgCOD/l
	X_{FS1}= nil

Design temperature is 20 ^{0}C and the necessary kinetic and stoichiometric parameters can be adopted from Table 3.2. The permissible effluent total COD limitation for this industrial wastewater category is 220 mgCOD/l. Testing has shown that sludge settling characteristics remain practically unaffected by fill time ratio.

2. *Select effective SRT* - Laboratory testing has shown that selection of an effective sludge retention time, θ_{XE} of 10 days is required for effective COD removal and reliable performance.

3. *Check effluent COD.* - Laboratory testing ensures that the selected effective SRT is sufficient for complete removal of influent biodegradable COD so that the effluent will only contain soluble inert COD of influent origin, S_{I1}, soluble inert metabolic products, S_P and particulate COD due to biomass escaping settling. S_{I1} of the industrial wastewater is determined as 100 mgCOD/l. S_P may be calculated from Equation 3.7, using f_{ES} = 0.07 experimentally determined for the wastewater:

$$SP = 0.07 \text{ x } 0.15 \text{ x } \frac{0.64}{1 + 0.15 \text{ x } 10} \text{ x } 1600 \text{ x } 13 = 57 \text{ mgCOD/l}$$

In this equation, the overall SRT is assumed as 13 days, to be confirmed in the following stages of design using Equation 2.10. The calculations confirm the results of laboratory testing indicating an effluent soluble residual COD, S_R of 160 mgCOD/l. Based on $i_{TSS,\,COD}$ = 0.9 mgTSS/mg COD given in Table 3.2, the permissible TSS concentration in the effluent, without violating the 220 mgCOD/l effluent COD limitation, is:

$$\text{Effluent TSS} = \frac{220 - 160}{0.90} = 67 \text{ mg TSS/l}$$

The SBR will consistently produce an effluent with appreciably lower TSS concentrations, thereby eliminating any need for additional measures to comply with the limitation.

4. *Calculate P_{XT}.* – The daily amount of sludge produced is calculated from Equations 3.8 and 3.9.

$$Y_{NH} = (1 + 0.2 * 0.15 * 10)\frac{0.64}{1 + 0.15 * 10} = 0.33 \text{cellCOD/COD}$$

$$P_{XT} = 0.9 * 1000(0.33 * 1600 + 200) = 660 kgTSS / d$$

5. *Calculate OR* – The daily amount of oxygen requirement is calculated from Equation 3.18.

$$OR_H = (1 - 0.33)1000 * 1600 = 1067 kgO_2 / d$$

6. *Select T_C* – The cycle time, T_C is selected as 6 h as a first approximation.
7. *Select $T_{S,}$ T_D and T_I* – A total period of 2 h is allocated for settle, draw and idle phases.
8. *Select SVI; calculate V_0/V_F* – An SVI level of 100 ml/g and a safety factor, SF of 1.3 are selected. The V_0/V_F ratio is then calculated from Equation 3.16:

$$\frac{V_0}{V_F} = 1.3 \left[0.9(0.33 * 1600 + 200)\ \right] \frac{10 * 24}{(6-2)}\ 100 * 10^{-6} = 5.1$$

The calculated value is off the usual range. Higher T_C values should be selected. Subsequent steps are continued for the initially selected T_C = 6 h but will be repeated for T_C values of 8, 12 and 24 h.

9. *Calculate HRT and V_T* – The hydraulic retention time, HRT is computed from Equation .2.3, which allows the calculation of the total volume requirement, $V_{T,}$ with fill volume and initial volume fractions:

$$\theta_h = (1 + 5.1)\, 6 = 36.6h$$

$$V_T = 1000 m^3 / d \frac{36.6h}{24h / d} = 1530 m^3$$

$$V_0 = 1280 m^3 \qquad V_F = 250 m^3$$

10. Calculate AOR and volumetric oxygen transfer rate - AOR and r_{OT} were calculated from Equations 3.19 and 3.21:

$$AOR = \frac{1067 kg / d}{24h / d(4/6)} = 66.6 kgO_2 / h$$

$$r_{OT} = \frac{(1 - 0.33)1600}{36.6(4/6)} = 43.5 gO_2 / m^3 h$$

or $$r_{OT} = \frac{66600 gO_2 / h}{1530 m^3} = 43.5 gO_2 / m^3 h$$

Using the assumption of 1.3kg AOR/kWh, the necessary surface aerator power may be computed as:

$$P = \frac{66.6 kgO_2 / h}{1.3 kgO_2 / kWh} = 51.2 kW$$

This simplifying assumption is specified not only to avoid calculation of the standard oxygen transfer rate, SOTR, but also the change in SOTR of surface aerators with respect to power input per unit volume. Since IEC standard electric motors are normally used by aerator manufacturers, one 55 kWaerator or two 30 kWaerators will be used. Checking for volumetric power input, (VPI), at V_0 and V_T respectively, indicates that both power input values remain at the admissible level.

$$VPI@0 = \frac{55000\,W}{1280m^3} = 43\,W/m^3 \text{ or } \qquad VPI@0 = \frac{60000\,W}{1280m^3} = 47\,W/m^3$$

$$VPI = \frac{55000\,W}{1530m^3} = 36\,W/m^3 \quad \text{or} \quad VPI = \frac{55000\,W}{1530m^3} = 39\,W/m^3$$

11. Select FTR- Since the treatment task in question is a carbon removal case and laboratory studies at this effective SRT indicate no restriction on the choice of reactor hydraulics, the FTR parameter is not regarded instrumental in reaching the treatment objectives. Priority will rather be given to minimizing the tankage requirement. Therefore, under diurnally uniform flow conditions fill time will be selected as T_C-T_{S+D} (i.e.T_P) for a single reactor resulting in varible FTR according to the selected T_C and $T_C/2$ for twin reactors resulting in FTR=0.5.

12. Calculate n and/or holding tank volume and the reactor dimensions- The total SBR volume of 1530 m3 can be built as a single reactor or two reactors each with half the volume.

i) Single reactor

A single reactor, 17*17*6.2 m (side water depth, SWD of 5.3m) will be built:

Freeboard	-	0.90
Fill	250	0.87
Safety supernatant	290	1.00
Settled sludge	990	3.43
Tank total	1530 m^3	6.20 m

During the 2 h settle and decant phases, the influent wastewater will have to be stored in a holding tank. Since 1000 m3 daily wastewater flow is uniform throughout the day in this case, active storage capacity of the equalization tank:

$$V_H = 1000m^3/d\frac{2h}{24h/d} = 83m^3$$

ii) Twin reactors

Two reactors, 13.5*13.5*5.0 (SWD of 4.2 m will be built:

Freeboard	-	0.80
Fill	125	0.69
Safety supernatant	145	0.80
Settled sludge	495	2.71
Tank total	765 m3	5.0 m

When two reactors are used no equalization is needed and only a pump pit or, in case of gravity inflow to the SBR, a flow diversion chamber will suffice.

The single or twin reactor alternatives above would have no marked difference in their total construction cost or land usage. As to associated equipment, a basic comparison is given below.

One reactor	Twin reactors
1 aerator, 55 kW	2 aerators, 30 kW each
1 decanter, 250 m3/h	2decanters, 125 m3/h each
1+1 equalization pump, 0.75kW	1 or 2 motor valves or penstocks

One reactor alternative would cost less in equipment, but its more significant advantage is greater simplicity in operation. It is obvious that three or more reactor systems would be even less competitive.

In the case of intermittent wastewater generation, selection of FTR should be changed. The concern being minimizing the total tankage cost, FTR should be selected as Q_{avg}/nQ_{max} for any number of reactor. Average flow, Q_{avg}, is computed as 1000/24=41.67 m^3/h and Q_{max} is computed as 1000/8=125m^3/h. This corresponds to a FRT of 1/3 for single reactor, 1/6 for twin reactors under 8 h/d flow condition. The minimum equalization volume requirement for each value of T_C can be computed as,

$$V_H = \delta \frac{Q_{avg} - Q_{min}}{Q_{avg}} Q_{max} T_F = 2 \frac{41.67 - 0}{41.67} 125 * 2 = 500 m^3$$

where δ is the number of occurrence of minimum wastewater flow cycles per day.

Results obtained for 4 different T_C values and for single or twin reactor alternatives under two different wastewater generation patterns are shown in Table 3.4. A scrutiny of the results in Table3.4 allows drawing the following conclusions:

V_0/V_F ratio remains within the usual 1-4 range for all values of T_C, except for T_C=6 h, where it climbs to an off-range value of 5.1. Since the wastewater is relatively strong, it is no surprise that selecting T_C=6h results in such a high recycle ratio. The data presented in the table clearly demonstrates that there is a critical cycle time which minimizes the SBR volume requirements. It is seen that this minimum is around T_C=8 h, corresponding to a HRT of 1.5 days. Greater and smaller values of T_C yield greater values of HRT, climbing for instance to around 2 days for T_C=24 h.

The necessary oxygen transfer rate per unit volume, r_{OT}, and the actual oxygenation rates, AOR, are observed to decrease with increasing T_C, reflecting the increase in aerated time fraction. Accordingly, the corresponding installed aerator power requirements decline from 51 kW to 37 kW. Volumetric power inputs at V_T remain in the allowable range, except for the value at T_C=24 h which represents typical mixing controlled conditions. Although the 37 kW aerator sized to deliver sufficient oxygenation could achieve complete mixing at V_0, it would be very insufficient at V_T. Higher power input would be needed to provide complete mixing. This would result in higher investment and running costs. Caution should still be exercised in using surface aerators in this case where V_0/V_F=0.94, since, with such variation of operating depth, modifications might have to be made in aerator and/or reactor construction design, as discussed in the previous section.

For 24h/d uniform flow conditions, T_C=8 h is the obvious choice. For the 8 h/d intermittent flow conditions T_C=12 h should be preferred since it combines minimum total tankage with minimum installed aeration power, considering the above discussed problems associated with T_C=24 h. On the other hand, it is still worthy to consider the choice of T_C = 24 h as it excludes holding tank requirements. Mixing will be needed in equalization tanks. The associated mixing and pumping equipment cost will be comparable to a 30 kW aerator cost. Not adhering to surface aerators , a better solution with T_C=24 h could be formulated. A 1935 m^3 single reactor with no equalization tank and associated equipment would be a superior choice in total investment cost, operating cost and operational simplicity.

Table 3.4 Summary of results for the design example

T_C (h)	V_0/V_F -	HRT (h)	AOR kg/h	r_{OT} gO2/m3.h	n*volume m3	n*P kW	VPI W/ m3	V_H 24h/d	V_H 8h/d
6	5.1	36.6	66	44	1x1530	1x55	36	84	500
					2x765	2x30	39	0	500
8	3.4	35.2	59	40	1x1475	1x45	31	84	667
					2x738	2x22	30	0	667
12	2.1	37.2	53	35	1x1530	1x45	29	84	500
					2x765	2x22	29	0	500
24	0.94	46.6	49	25	1x1935	1x37.5	19	84	0
					2x968	2x18.5	19	0	0

4

Process Design for Nutrient Removal

4.1. BASIC PRINCIPLES FOR NITROGEN REMOVAL

Biological nitrogen removal from wastewaters is the most feasible scheme among a number of alternative physical and chemical processes. Domestic sewage and the majority of industrial wastewaters do not generally contain oxidized nitrogen; the total nitrogen content, (C_{TKN1}), of domestic sewage is usually in the range of 40 to 60 mg N/l, with 30 to 40 mg N/l of ammonia nitrogen and 10 to 20 mg/l of organic nitrogen (Orhon *et al.,* 1997; Rössie and Pretorius, 2001). Therefore, biological nitrogen removal proceeds as a sequence of two different processes, first *nitrification* which oxidizes ammonia to nitrite, (NO_2^-) and nitrate, (NO_3^-), and then, *denitrification* which reduces nitrate and returns it to the atmosphere as molecular nitrogen, (N_2). The relevant biochemical reactions that occur in single sludge processes may be summarized as:

- The microbial degradation of nitrogen-containing compounds and the release of ammonia nitrogen, S_{NH}.
- The incorporation of nitrogen into cells by biosynthesis and its removal by excess sludge
- The microbially mediated redox transformations, including nitrification and denitrification.

Biological conversion of organic nitrogen to ammonia takes place through hydrolysis of the particulate organic nitrogen fraction into simpler soluble compounds, which then undergo ammonification reactions (Dold and Marais, 1986; Henze *et al.,* 1987). For domestic sewage, particulate organic nitrogen is usually a small fraction of the total organic content and all the complex nitrogen forms can be readily deaminated for the range of SRT values commonly adopted for the SBR operation. In practice, the influent total nitrogen content, C_{TKN1} with the exception of non-biodegradable nitrogen components may be approximated as ammonia nitrogen for system design.

Process stoichiometry for nitrogen removal should account for the amount of nitrogen incorporated into active biomass through biosynthesis reactions. The COD/N ratio of the domestic sewage is well below that in activated sludge so that only 20-30% N removal can be attributed to microbial growth.

Ammonia oxidation through nitrification is the decisive step controlling the magnitude of the nitrogen removal performance. The autotrophic yield coefficient, Y_A is the essential parameter determining the stoichiometry of nitrification. Based on theoretical considerations and observations oxygen requirements for nitrate formation, a Y_A value of 0.24 g cell COD/g N is commonly accepted for design evaluations (Henze *et al.,* 1987). A rational design approach should also take into account that the maximum specific growth rate for nitrifiers, μ_{Amax}, is much lower than that of the heterotrophs, and its value is seriously affected by environmental factors such as the dissolved oxygen, temperature and pH of the growth medium. The extent of nitrification and the level of the ammonia concentration in the reactor are determined by the growth kinetics of the nitrifiers, represented by the following Monod expression:

$$\mu_A = \mu_{A\max} \frac{S_{NH}}{K_{NH} + S_{NH}} \tag{4.1}$$

The level of μ_A in the above expression is quite critical as it determines, together with the endogenous respiration rate for nitrifiers, b_A, the minimum aerobic SRT likely to cause washout of the nitrifying biomass from the reactor.

In denitrification, nitrate or nitrite replace under anoxic conditions dissolved oxygen in microbial respiration. Denitrification and organic carbon removal are assumed to rely on the same heterotrophic biomass, except for the fact that they use different electron acceptors. They must be evaluated however in terms of opposite driving forces involved for design: In systems designed solely for organic carbon removal, the required amount of electron acceptor should be calculated and supplied for complete utilization of the biodegradable COD; For systems involving denitrification, however, the main objective is to remove nitrate which is the electron acceptor, and a sufficient amount of biodegradable COD should be present for the reduction of the available nitrate in the reactor.

A factor of 2.86 g O_H/g NO_3^-N sets the equivalence between COD and nitrate nitrogen.

In single-sludge activated sludge systems operated for nitrogen removal, heterotrophic biomass using dissolved oxygen is expected to readily adapt itself to start consuming nitrate under anoxic conditions. In this cyclic operation, the observed rate of substrate utilization in the anoxic phase is always lower than what is normally achieved under aerobic conditions (Henze *et al.,* 1987; Orhon and Artan, 1994; Orhon *et al.,* 1996). In system design and modelling, this observation is accounted for simply by adopting an empirical coefficient $\eta < 1.0$, which acts as an overall correction factor for anoxic conditions.

4.1.1. Nitrogen Mass Balances

Nitrogen removal performance in SBR systems, as in all activated sludge configurations, depends upon the balance between the three key parameters, namely the nitrification capacity, N_{OX}, the denitrification potential, N_{DP}, and the available nitrate or the magnitude of oxidized nitrogen supplied to the mixed periods, N_A. All three parameters are conveniently defined in mass per unit volume of wastewater, allowing simple manipulation in mass balance. The nitrification capacity, N_{OX} may be calculated from the basic mass balance for ammonia nitrogen:

$$N_{OX} = C_{TKN1} - i_{NXI}X_{I1} - i_{NSI}S_{I1} - N_X - S_{NH} \quad (4.2)$$

In this expression, the non-biodegradable nitrogen components, by-passing biological conversion are defined as fractions (i_{NSI}; i_{NXI}) of their soluble and particulate COD counterparts in the wastewater, S_{I1} and X_{I1}. The effluent ammonia nitrogen concentration, S_{NH} is a function the aerobic SRT.

The net heterotrophic yield coefficient, $Y_{NH,}$ which is the function of θ_{XE}, determines the part of the nitrogen concentration that is incorporated into biomass and removed from the system as part of the excess sludge, N_X. Calculation of N_X requires the assessment of nitrogen content of biomass, i_{NBM}:

$$N_X = i_{NBM} Y_{NH} C_{S1} \quad (4.3)$$

The denitrification potential, N_{DP} indicates the concentration of nitrate nitrogen that may be potentially removed, provided that enough nitrate nitrogen is supplied to the non-aerated period. It can be defined as a fraction of the nitrate nitrogen equivalent of the total electron acceptor demand associated with the heterotrophic growth on both readily biodegradable substrate, N_{SS}, and slowly biodegradable substrate, N_{XS}, and endogenous respiration, N_{ER}.: The expressions assume that all biodegradable COD is consumed for the selected effective SRT:

$$N_{SS} = (1 - Y_H)\frac{S_{S1}}{2.86} \quad (4.4)$$

$$N_{XS} = (1 - Y_H)\frac{X_{S1}}{2.86} \quad (4.5)$$

$$N_{ER} = (1 - f_E) b_H \theta_{XE} \frac{Y_H}{1 + b_H \theta_{XE}} \frac{(S_{S1} + X_{S1})}{2.86} \quad (4.6)$$

Neglecting the concentration gradient of the growth limiting substrate throughout the anoxic and aerobic periods, an acceptable assumption for intermittent aeration and simultaneous denitrification, the available fractions of N_{SS}, N_{XS} and N_{ER} will be directly related to the T_{AOX}/T_E ratio and hence total denitrification potential of the system, N_{DP} will be defined as follows:

$$N_{DP} = \frac{T_{AOX}}{T_E}(\eta_g N_{SS} + \eta_h N_{XS} + \eta_e N_{ER}) \quad (4.7)$$

where η is the reduction factor for process rates under anoxic conditions. However, in an SBR operation with an anoxic/aerobic sequence in the cycle, (predenitrification), it will be justifiable to assume complete depletion of the readily biodegradable COD within the anoxic period provided that filling takes place only during a portion of that period. Then, the above expression may be arranged to yield the denitrification potential of the SBR operation in a *predenitrification* mode:

$$N_{DP} = N_{SS} + \frac{T_{AOX}}{T_E}(\eta_h N_{XS} + \eta_e N_{ER}) \quad (4.8)$$

Oxidized nitrogen is supplied to the first anoxic period by recycle and to the following anoxic periods from the preceding aerobic period. In an SBR operation with a single anoxic/aerobic sequence, (predenitrification), the recycle ratio, V_0/V_F, is the key parameter that sets the magnitude of nitrate nitrogen that will be introduced into the anoxic period:

$$N_A = \frac{V_0}{V_F} S_{NO} \tag{4.9}$$

where, S_{NO} denotes the effluent nitrate nitrogen concentration. In the case of multiple sequence of mixing/ aeration phases within a cycle, however, recycle supplies nitrate only to the first anoxic period while the others utilize the nitrogen oxidized in the preceding aerobic periods. Hence, in these SBR systems, the total available nitrate is not determined solely by the V_0/V_F ratio. So, both N_{DP} and N_A depend on selected process options i.e. aeration and filling strategies employed.

4.1.2. Carbon and Nitrate Limitations

The denitrification efficiency of a process is limited by either the denitrification potential, N_{DP}, or the available oxidized nitrogen, N_A. In the case of N_{DP} limitation, ($N_{DP}<N_A$), the effluent nitrate concentration will be determined by the difference between N_{OX} and N_{DP}:

$$S_{NO} = N_{OX} - N_{DP} \tag{4.10}$$

With this basic definition, the denitrification efficiency, E_P, potentially achievable through N_{DP} may be expressed as follows:

$$E_P = N_{DP} / N_{OX} \tag{4.11}$$

It should be underlined that the potential denitrification efficiency, E_P, is directly related to COD/TKN ratio of the wastewater for a selected effective SRT. An evaluation example is provided as outlined in Table 4.1 to visualize the effect of the COD/TKN ratio and the effective SRT on E_P. The evaluation involves two different wastewaters characterized by total COD levels, (C_{T1}), of 300 mg/l and 600 mg/l, and the same total biodegradable nitrogen content, (C_{TKN1}), of 50 mg N/l, thus corresponding to COD/N ratios of 6 and 12, respectively. The same total biodegradable COD fraction, f_{CS} = 0.80, the readily biodegradable fraction, f_{SS} = 0.20 and η = 0.8 were adopted for the two wastewaters. Other coefficients were taken, as required, from Table 3.2. An aerobic sludge age value of θ_{XA}= 6 d, ensuring complete nitrification was adopted as the constant design parameter. The effective SRT was calculated from Equation 2.14 as a function of the selected T_{AOX}/T_E ratio. The exercise basically compared the *predenitrification* mode of operation enabling full use of the readily biodegradable COD of the wastewater as indicated in Equation (4.8) with a more general *multiple mixing/aeration* scheme associated with an N_{DP} defined in Equation (4.7). Consequently, N_{DP} were alternately calculated from Equations (4.7) and (4.8) for the two wastewaters associated with COD/N ratios of 6 and 12.

Table 4.1 essentially indicates that the *predenitrification* mode has the ability to provide a higher denitrification efficiency for a given anoxic fraction, especially for systems operating at low anoxic fractions. As can be seen from the table, a denitrifiction efficiency of over 75% can not be reached for wastewaters with a relatively low COD/TKN ratio of 6, using the highest acceptable anoxic fraction. A N_{DP} value, lower than what needs to be removed through denitrification for the highest acceptable limit for the T_{AOX}/T_E value (generally T_{AOX}/T_E < 0.60), is a decisive indication that internal carbon is not sufficient enough to provide the required nitrogen removal for the particular wastewater. If the COD/TKN ratio is higher (600/50=12) both operation modes have the potential to remove all oxidized nitrogen with a relatively low anoxic fraction (0.45 for multiple aeration; 0.30 for predenitrification). The denitrification efficiency values greater than one ($E_P > 1.0$) in Table 4.1 show that a fraction of the denitrification potential will be wasted since available nitrate is limited, i.e. the selected anoxic fraction (in fact, mixed fraction) is too high for nitrogen removal alone.

Table 4.1. Effect of the COD/TKN ratio, the anoxic fraction and the aeration pattern on potential denitrification efficiency

	multiple aeration		predenitrification	
C_{T1} (mgCOD/l)	300	600	300	600
T_{AOX}/T_E	N_{DP}/N_{OX}	N_{DP}/N_{OX}	N_{DP}/N_{OX}	N_{DP}/N_{OX}
0.1	0.10	0.22	0.27	0.62
0.2	0.20	0.45	0.35	0.81
0.3	0.30	0.69	0.44	1.01
0.4	0.41	0.93	0.54	1.22
0.5	0.53	1.19	0.64	1.43
0.6	0.65	1.45	0.75	1.66

In the case of available nitrate limitation, ($N_A<N_{DP}$), the effluent nitrate concentration will be determined by the difference between N_{OX} and N_A:

$$S_{NO} = N_{OX} - N_A \tag{4.12}$$

Introducing the definition of N_A for *predenitrfication* systems in Equation (4.9) into the above expression, the effluent nitrate concentration and hence the denitrification efficiency can be defined as a function of V_0/V_F:

$$S_{NO} = \frac{N_{OX}}{1 + V_0/V_F} \tag{4.13}$$

$$E = \frac{N_A}{N_{OX}} = \frac{V_0/V_F}{1 + V_0/V_F} \tag{4.14}$$

The above expression shows that denitrification efficiency higher than 80% can not be expected from predenitrification systems if the maximum limit adopted for the recycle ratio is $V_0/V_F = 4$. In the case of multiple mixing/aeration, the same denitrification potential can be attained with a higher anoxic fraction, as can be seen from Table 4.1. In systems with multiple anoxic phases, however, lower effluent nitrate concentrations can be achieved compared to predenitrification, due to higher levels of N_A, since this parameter is not solely determined by the recycle ratio, if the COD/N ratio of the wastewater is appropriate. The positive difference between N_{DP} and N_A values of a system ($N_{DP} > N_A$) can also be interpreted as an index for the possibility of EBPR.

4.1.3. Selection of Process Option

Wastewater characteristics and effluent requirements are the most important factors affecting process selection. Regulations are normally set in terms of the allowable total nitrogen concentration, $C_{N,E}$ in the effluent. Based on this effluent limit, a design effluent nitrate nitrogen concentration, $S_{NO,D}$ that would safely provide compliance with regulatory restrictions may be calculated, considering the remaining ammonia concentration, the inert fraction of the soluble organic nitrogen and the nitrogen fraction of the biomass escaping in the effluent:

$$S_{NO,D} = C_{N,E} - S_{NH,E} - i_{NSI}S_{II} - i_{N\,VSS}X_{VSS,E} \tag{4.15}$$

Assessment of an appropriate value for $S_{NO,D}$ determines $N_{OX} - S_{NO,D}$, the fraction of the oxidized nitrate concentration that needs to be removed through denitrification. The condition below should be satisfied as an indication that the desired process efficiency is potentially achievable.

$$N_{DP} \geq N_{OX} - S_{NO,\,D} \tag{4.16}$$

If not, the selected anoxic fraction should be increased to the acceptable high limit to provide additional denitrification. In case when N_{DP} is not limiting, the basic question remains to be able to supply to the anoxic zone, the sufficient amount of nitrate to be removed, N_A. Recycle is the instrument for the S_{NO} supply in predenitrification systems with the following mass balance:

$$N_{OX} - S_{NO,D} = N_A = \frac{V_0 / V_F}{1 + V_0 / V_F} N_{OX} \tag{4.17}$$

Since predenitrification yields the highest denitrification potential for the same anoxic fraction, it should be preferentially selected for wastewaters having a low COD/TKN ratio. If the required denitrification efficiency is below 80 % (since SBR operation imposes a maximum limit of $V_0/V_F = 4$) this operation mode may be the appropriate option; the problem is to select a T_C value that would yield the required recycle ratio (V_0/V_F) for desired nitrogen removal. V_0/V_F calculated as given in Equation (3.16) only relates to biomass in the reactor and settling properties, and approximates in a way, the sludge recycle in continuous-flow systems. The value of V_0/V_F may be increased, as indicated by this equation, by selecting a smaller value of T_C. In SBR systems for nutrient removal, however, the required recycle ratio for the desired nitrogen removal efficiency may not be achieved even if T_C is selected as short as a lowest acceptable limit (for example 4h) for weak wastewaters. In order to obtain the required nitrogen removal, in this case, V_0/V_F has to be increased to provide the corresponding nitrate recycle which will inevitably yield an unnecessarily high HRT. If this solution is not acceptable, a *step feeding* type of operation mode can be the appropriate solution. Since denitrification efficiency is not solely determined by the recycle ratio in this process, a lower V_0/V_F ratio associated with an optimum T_C value can be used. For strong wastewaters, on the other hand, optimum T_C values may already yield higher V_0/V_F ratios as discussed in previous chapter. Therefore, *predenitrification*, aside being the best option for wastewaters with relatively low COD/TKN ratios (for example COD/TKN<7), should also be considered as the suitable process alternative for wastewaters with high COD concentrations ($C_{S1} \geq$ 1000 mg/l) when the required denitrification efficiency is lower than 80%.

If the required denitrification efficiency is higher than 80% as a consequence of treating a strong nitrogenous wastewaters or complying with very low effluent nitrogen standards, this can be accomplished by also using a *multiple anoxic/aerobic* operation mode, provided that COD/TKN ratio is high enough.

4.2. UNIFIED DESIGN PROCEDURE FOR NITROGEN REMOVAL

This section describes a unified design procedure for SBR systems operated for nitrogen removal. The design of nitrogen removing SBR systems is structured upon nitrogen mass balances and benefits from all of the principles and procedures presented in Chapter 3. The proposed procedure is schematically described in Figure (4.1) and further explained by illustrative design examples.

The starting point of SBR design for nitrogen removal is the selection of an appropriate aerobic sludge retention time, θ_{XA}, that will ensure nitrification. The efficiency of nitrogen removal depends upon the extent of denitrification achieved under employed effective sludge retention time, θ_{XE}. Appropriate design should also ensure provision of the necessary amount of nitrate for denitrification in the anoxic periods.

The selected aerobic SRT should be greater by a safety factor, than a minimum aerobic sludge age at which all nitrifiers will be washed out of the systems. Neglecting decay of nitrifiers in the unaerated periods, the following mass balance defines the relationship between θ_{XA} and the maximum autotrophic growth rate, $\mu_{Amax,}$ which should be determined for the particular wastewater and accounting for the significant process conditions such as temperature, DO concentration, pH, etc.

$$\frac{1}{\theta_{XA}}=\mu_{A\max}\frac{S_{NH}}{K_{NH}+S_{NH}}-b_A \tag{4.18}$$

Upon selection of the aerobic SRT, effluent ammonia nitrogen concentration, S_{NH} can be estimated from the equation below:

$$S_{NH}=\frac{K_{NH}(1+b_A\theta_{XA})}{\mu_{A\max}\theta_{XA}-(1+b_A\theta_{XA})} \tag{4.19}$$

It should be noted that the above expression only gives a rough and conservative approximation for SBR since it reflects mass balance for a continuous-flow, completely mixed reactor.

The anoxic fraction of the effective process time, (T_{AOX}/T_E), must then be selected to provide the necessary N_{DP} satisfying the condition given in Equation (4.16). Determination of the anoxic fraction allows calculation of the effective SRT, θ_{XE} and hence N_{DP}. The other design parameters, namely, N_X, N_{OX}, the sludge production and the oxygen requirement may then be calculated in terms of the effective sludge retention time. Adopting *predenitrification* as the initial option, N_{DP} is calculated from Equation (4.8) to check that internal carbon is sufficient enough to provide the required nitrate removal. Iteration may be continued, if necessary, by increasing T_{AOX}/T_E values up to the acceptable limit ($T_{AOX}/T_E < 0.6$).

Sludge production and oxygen requirement equations given in Chapter 3 must be revised considering nitrification and denitrification. Heterotrophic biomass will not change if yield and decay coefficients under aerobic and anoxic conditions are assumed to be the same. Contribution of the autotrophic biomass to the sludge production is included in the following expression:

$$P_{XT}/Q=i_{TSS,COD}\left(Y_{NH}C_{SI}+Y_{NA}N_{OX}+X_{I1}\right)+X_{FSI} \tag{4.20}$$

where Y_{NA} is the net autotrophic yield coefficient which can be defined as follows as a function of the effective SRT:

$$Y_{NA}=(1+f_E b_A\theta_{XE})\frac{Y_A}{1+b_A\theta_{XE}} \tag{4.21}$$

A substantial increase has to be considered, as indicated in the expression below, in the calculation of the oxygen requirement due to additional oxygen consumption for nitrification, partly compensated by denitrification:

$$OR_T / Q = (1 - Y_{NH})C_{SI} + (4.57 - Y_{NA})N_{OX} - 2.86(N_{OX} - S_{NO}) \quad (4.22)$$

As shown in Figure (4.1), the next step in the design procedure is to verify recycling requirement. With the provision that there is no organic carbon limitation ($N_{DP}>N_A$), the required recycle ratio, R, which would yield the desired effluent nitrate concentration, can be calculated from Equation (4.17). If R > 4, the predenitrification option is not feasible other process alternatives should be considered. If the predenitification scheme is acceptable (R < 4), a suitable value for T_C should be selected so as to yield a V_0/V_F ratio from Equation (3.16) that would satisfy recycle requirements for nitrogen removal providing the necessary available nitrate, N_A for the following anoxic period.

In addition to the considerations in the previous chapter (reactor volume and aeration capacity optimization) nitrate recycle is an important factor in the selection of the cycle time. Iteration should be started from a lowest acceptable T_C value corresponding to the highest possible V_0/V_F ratio. For weak wastewaters, however, optimum T_C may not yield the required recycle ratio (V_0/V_F < R). Then, other process options should be selected and N_{DP} prediction should be revised in accordance with the selected filling and mixing/aeration pattern.

4.2.1. SBR Design for Pre-denitrification

A design exercise for nitrogen removal by means of predenitrification is provided in this section using three types of wastewaters with different COD and nitrogen contents – and consequently different COD/TKN ratios - for comparative evaluation. The first one (WW1) was selected to reflect the average character of Istanbul sewage with a COD/TKN ratio of around 9.5 (Orhon *et al.*1998). One of the other two strong wastewaters (WW2) represents a mixture of leachate and sewage (Diamanopoulos *et al.,* 1997). It has a higher TKN content with a somewhat lower COD/TKN ratio of 8.1 compared to domestic sewage; it should be noted that the available organic carbon for denitrification is in reality much lower than what is indicated by this ratio if it is calculated on a biodegradable COD basis. The other one (WW3) is a plain-settled tannery effluent, much stronger both with respect to COD and TKN contents and with a resulting COD/TKN ratio of around 14 (Murat *et al.,* 2002). Characteristics of the wastewaters for design are outlined in Table 4.2 together with selected kinetic and stoichiometric coefficients for autotrophs. Other necessary coefficients were adopted from Table 3.2. Effluent concentrations for design are selected as $S_{NH,D}$=2mgN/l and $S_{NO,D}$=10mgN/l to allow reliable compliance with relevant discharge requirements.

The proposed design procedure was implemented for the selected wastewaters. As illustrated in Figure 4.1, the starting step for the design is the selection of the aerobic sludge age, θ_{XA}. This parameter was selected as 8 days for the domestic wastewater and 10 days for the others due to possible inhibitory effects associated with industrial wastewaters. Effluent ammonia concentration, S_{NH} was then calculated as 1.4 mg/l for the domestic sewage and 1.5 mg/l for the other wastewaters using Equation 4.19 and kinetic coefficients given in Table 4.2: As the adopted value for $S_{NH,\ D}$ is 2.0 mg/l, $S_{NH} < S_{NH,\ D}$ condition is safely satisfied and confirms that θ_{XA} selection is acceptable.

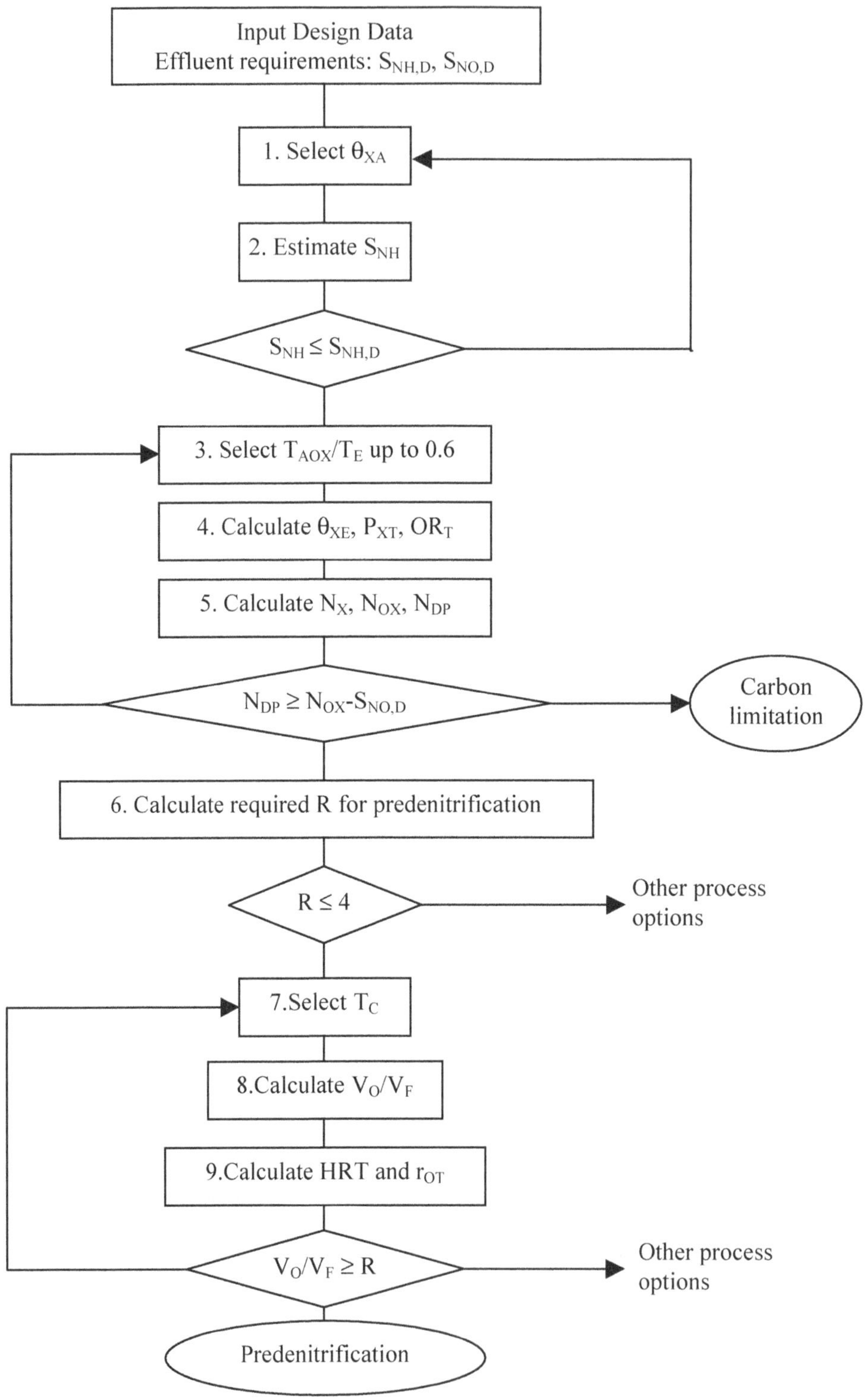

Figure 4.1.The design algorithm for nitrogen removal

The next step in the design procedure is the selection of an appropriate T_{AOX}/T_E ratio. A T_{AOX}/T_E value of 0.38 is estimated by iteration for the domestic sewage. This value yields an effective sludge age, θ_{XE} of 12.9 days and enables to calculate the net heterotrophic yield, Y_{NH} as 0.30 mg cell COD/mg COD and nitrogen removed through incorporation into biomass, N_X as 11.5 mgN/l. With this information, the two major design parameters, the nitrification capacity, N_{OX} and the denitrification potential, N_{DP} are then computed as 42 mgN/l and 33 mgN/l respectively. With these data, the design evaluation indicates that the selected effluent nitrate concentration, $S_{NO,D}$ of 10 mg/l can be secured, as $N_{DP} > N_{OX} - S_{NO,D}$ condition is satisfied. This performance however, can only be achieved if a minimum of 32 mgN/l nitrate is recycled into the anoxic phase (N_A = 32 mg/l). A recycle ratio, R of 3.2 < 4.0 provides the necessary N_A to be denitrified during the anoxic phase. Based on available data, other necessary parameters such as sludge production and oxygen requirement may also be computed as given in Table 4.3.

Table 4.2 Selected influent wastewater characteristics.

Parameter	Symbol	WW1	WW2	WW3	Unit
Total COD	C_{TI}	525	1090	2330	$mgCODl^{-1}$
Biodegradable COD	C_{SI}	450	600	2000	$mgCODl^{-1}$
Readily biodeg. COD	S_{SI}	52	40	460	$mgCODl^{-1}$
Inert. soluble COD	S_{II}	25	160	65	$mgCODl^{-1}$
Inert Particulate COD	X_{II}	50	330	265	$mgCODl^{-1}$
Total Kjeldahl Nitrogen	C_{TKNI}	55	133	167	$mgNl^{-1}$
N content of biomass	i_{NBM}	0.085	0.085	0.085	N/COD
N content of X_I	i_{NXI}	0.01	0.01	0.05	N/COD
N content of S_I	i_{NSI}	0.03	0.03	0.03	N/COD
Fixed Solids	X_{FSI}	0	0	0	$mgTSSl^{-1}$
Autotrophic yield	Y_A	0.24	0.24	0.24	COD/N
Autotrophic endogenous decay coefficient	b_A	0.05	0.05	0.05	day^{-1}
Half saturation coefficient for nitrification	K_{NH}	1.0	1.0	1.0	$mgNl^{-1}$
Maximum growth rate for autotrophs	$\hat{\mu}_A$	0.30	0.25	0.25	day^{-1}
Correction factor for growth	η_g	0.8	0.8	0.8	-
Correction factor for hydrolysis	η_h	0.6	0.8	0.8	-

The design exercise outlined in Table 4.3 indicates that the leachate-domestic wastewater mixture (WW2) is a typical example of a strong nitrogenous wastewater with an internal carbon pool significantly lower than what is required to achieve the desired N removal efficiency. A T_{AOX}/T_E value of 0.60, the highest acceptable anoxic fraction, yields at the limit, an N_{DP} level of 80 mgN/l, representing only 68% of the N_{OX} associated with the selected operation conditions. In other words, the maximum denitrification efficiency cannot be higher than 68%, corresponding to an effluent nitrate nitrogen concentration, S_{NO} of 37 mg/l. Consequently, the desired effluent S_{NO} level cannot be achieved for this wastewater unless external carbon is used. If a predenitrification process is adopted without carbon supplement, a recycle ratio of 2.1 is sufficient to utilize the full denitrification potential of the system.

Table 4.3 Selected design parameters and results derived from the optimum θ_{XE}

Parameter	Symbol	Unit	WW1	WW2	WW3
Aerobic sludge age	θ_{XA}	day	8	10	10
Effluent ammonium nitrogen	S_{NH}	mgN/l	1.4	1.5	1.5
Anoxic sludge fraction	T_{AOX}/T_E	-	0.38	0.60	0.11
Effective sludge age	θ_{XE}	day	12.9	25	11.2
Net heterotrophic yield	Y_{NH}	COD/COD	0.30	0.23	0.32
Nitrogen in biomass	N_X	mgN/l	11.5	11.8	53
Nitrification capacity	N_{OX}	mgN/l	42	117	96
Denitrification potential	N_{DP}	mgN/l	33	80	86
Required available nitrate	N_A	mgN/l	32	80	86
Denitrification efficiency	E	%	76	68	90
Required recycle ratio	R	-	3.2	2.1	9
Effluent nitrate nitrogen	S_{NO}	mgN/l	10	37	19*
Sludge production	P_{XT}/Q	mgTSS/l	173	450	830
Total oxygen requirement	OR_T/Q	mgO_2/l	412	735	1590

* calculated as %20 of N_{OX}

The tannery effluent (WW3) is a strong wastewater with adequate internal carbon satisfying the required N_{DP} level with a relatively low anoxic fraction of only 0.11. However the magnitude of the oxidized nitrogen (N_{OX} = 96 mg/l) is too high for providing the desired effluent nitrate concentration of 10 mgN/l; it requires in principle an N_A level of 86 mgN/l to be recycled into the anoxic period, and a denitrification efficiency of around 90%, only achievable with a recycle ratio of as high as 9. Therefore, the system becomes R limiting and can only provide an S_{NO} value of 19 mg/l for the limit R value of 4.0.

The next step in the design procedure is to choose the optimum cycle time, T_C, which will yield the nitrate recycle ratio specifically required for each type of wastewater. Table 4.4 shows the values of the recycle ratio, V_0/V_F, the hydraulic retention time, HRT and the volumetric oxygen transfer rate, r_{OT} for a selected range of T_C. For domestic sewage, a T_C of 4 hours, the lowest in the range considered, appears to be the only cycle time that yields a V_0/V_F of 3.3, slightly higher than the required recycle ratio. A T_C of 4 hours allows an effective time, T_E of 2 h and hence a T_{AOX} of 0.8 h. Therefore the selected fill phase, T_F must be equal or lower than mixed period for predenitrification ($T_F \leq 0.8$ h). Table 4.4 shows that T_C of 4 h results in a slightly higher HRT and higher oxygen transfer rate values compared to the other options. So, a *step-feeding* alternative with a higher T_C value should be evaluated as a more feasible design option.

For the leachate-domestic wastewater mixture (WW2), the optimum cycle time, T_C is 12 h, yielding aside a more than sufficient nitrate recycle ratio, the lowest HRT and relatively smaller aeration equipment. It should be noted that the higher V_0/V_F value of 3.5 that results from the selected TC does not in any way improve the denitrification efficiency, as the system is N_{DP} (carbon) limited. The constraint of $T_F \leq T_{AOX}$ ($T_F < 6$ h) must also be observed.

A cycle time of 8 hours, which yields a recycle ratio of 4.9 (slightly higher than 4.0) and hence 16 mg/l effluent nitrate, should be selected for tannery wastewater. Lower T_C values will lead to V_0/V_F ratios beyond the practically acceptable range. Other process options, such as *intermittent aeration* can be tested to improve the denitrification efficiency.

Table 4.4 Selected TC values and their effects on design

Domestic Wastewater				Domestic & Leachate				Tannery wastewater			
T_C (h)	$\frac{V_0}{V_F}$	HRT (h)	r_{OT} g/m^3.h	T_C (h)	$\frac{V_0}{V_F}$	HRT (h)	r_{OT} g/m^3.h	T_C (h)	$\frac{V_0}{V_F}$	HRT (h)	r_{OT} g/m^3.h
4	3.3	17	75	10	4.4	54	43	6	7.4	50	54
6	1.7	16	59	12	3.5	54	41	8	4.9	47	51
8	1.1	17	50	24	1.6	62	32	12	2.9	47	46

Predenitrification systems mostly yield effluent nitrate concentration somewhat lower than the predicted one from Equation 4.13 likely due to simultaneous nitrification/denitrification, (SND) taking place to a certain extent during the aerated period of the cycle. Anoxic zone may occur in the floc depending on the oxygen concentration in the bulk liquid. The magnitude of SND is determined by the dissolved oxygen concentration and the oxygen half saturation constant, K_O for heterotrophic biomass. This issue will be discussed further in the following chapter. In the case of predenitrification mode with simultaneous nitrification/denitrification in the aerated period, the mass balance Equation 4.13, defining the effluent nitrate concentration, S_{NO} should be modified as below, to account for the amount of nitrogen denitrified in aerated period, N_{SND}:

$$S_{NO} = \frac{N_{OX} - N_{SND}}{1 + V_0/V_F} \tag{4.23}$$

4.2.2. SBR Design for Step Feeding

An SBR operation with dual mixed filling basically requires a second set of mixed and aerobic periods, T_{A2} and T_{AOX2} within the same cycle (see Figure 2.3.(b)). A β fraction of the total fill volume (V_{F2}=βV_F), is diverted to the second anoxic period to create the necessary denitrification potential:

$$N_{DP2} = \beta N_{SS} + \eta \frac{T_{AOX2}}{T_E}(\beta N_{XS} + N_{ER}) \tag{4.24}$$

Similarly, denitrification potential of the first anoxic reactor, N_{DP1} can be expressed as:

$$N_{DP1} = (1-\beta)N_{SS} + \eta\left(\frac{T_{AOX}}{T_E} - \frac{T_{AOX2}}{T_E}\right)[(1-\beta)N_{XS} + N_{ER}] \tag{4.25}$$

The nitrate nitrogen generated in the first aerobic period, N_{A1} will be available for the second anoxic period and this way, additional nitrate removal can be achieved by post denitrification. Assuming that the ammonia load introduced into the first stage is entirely oxidized in this stage, N_{A1} will be proportional to (1- β). In order to achieve complete ammonias oxidation in each stage, fractionation of total aerobic period should be done in proportion to the ammonia load introduced into each anoxic period since it will be converted into nitrate at the following aerobic period; thus the fraction of the first aerobic period (T_{A1}/T_A) should best be selected to approximate (1- β).

$$N_{A1} = (1-\beta)N_{OX} \tag{4.26}$$

Provided that the denitrification potential of the second anoxic period, N_{DP2} is sufficient to remove all nitrate introduced from the first stage (N_{A1}), oxidized nitrate in the second stage, βN_{OX} will partly recycle to the first anoxic period with sludge, in proportion with V_0/V_F and will provide the available nitrate, N_{A2} for denitrification in the first stage:

$$N_{A2} = \frac{V_0/V_F}{1+V_0/V_F}\beta N_{OX} \tag{4.27}$$

The remaining portion of βN_{OX} will leave the system as effluent nitrate nitrogen:

$$S_{NO} = \frac{\beta N_{OX}}{1+V_0/V_F} \tag{4.28}$$

For a given total anoxic fraction (T_{AOX}/T_E), optimum β and T_{AOX2}/T_E values to obtain minimum effluent nitrate concentration can be calculated using the following two key nitrogen balances to be satisfied around each sub-cycle:

$$N_{DP2} = N_{A1} \tag{4.29}$$

$$N_{DP1} = N_{A2} \tag{4.30}$$

A lower S_{NO} value requires lower β on the basis of mass balance defined in Equation 4.28 and higher N_{DP}, which can only be obtained by increasing the total anoxic fraction. This will lead to increased second anoxic fraction allowing very small fraction for first anoxic period. If total anoxic fraction, T_{AOX}/T_E, is further increased, more than two anoxic/aerobic sub-cycles are required in order to fully utilize the portion of the denitrification potential in the first anoxic period since the available nitrate for pre-denitrification will be limited. In this case optimizing the system operation with a triple anoxic/aerobic period may solve the problem, with the assumption that remaining fill volume fraction (1-β) is equally subdivided between the first two anoxic periods.

A set of three different design exercises is outlined below to illustrate the rationale behind the preferential selection of a *step-feeding operation* for nitrogen removal. The domestic sewage characteristics previously defined in Table 4.2 for predenitrification, (WW1), are used again to allow comparative evaluation. In the first exercise, *(Dual 1)*, the same values of 10 mgN/l for the effluent nitrate concentration and 8 days for θ_{XA} are adopted for the evaluation. The design iteration is started with $T_{AOX}/T_E = 0.40$, a value slightly higher than that of predenitrification since a portion of the potentially available N_{DP} will inevitably be wasted in the first aerobic period. Then, a T_C level of 8 h is selected, yielding a V_0/V_F ratio of 1.2. In the next step, optimum β and T_{AOX2}/T_E values are calculated using the nitrogen balances given in Equations 4.29 and 4.30, for the initially adopted T_{AOX}/T_E value. The equations give $\beta = 0.49$ and $T_{AOX1}/T_E = 0.12$, leading to an S_{NO} value of 9.4 mgN/l satisfying the adopted effluent limitation. The other design parameters are outlined in Table 4.5. This example illustrates that a low T_C value, a major handicap in the practical operation of a *predenitrification* scheme, is no longer a requirement for *step feeding*, because a significant fraction of nitrate is removed in the second anoxic period and therefore a high V_0/V_F is not needed. Moreover freedom from $V_0/V_F < 4.0$ restriction allows further decreasing the S_{NO} levels at the expense of a higher T_{AOX}/T_E ratio and a lower β value. The second design example (Dual 2) illustrates a case where a lower $S_{NO,\ D}$ level of 7.0 mgN/l is imposed. Accordingly, selecting a higher T_{AOX}/T_E ratio of 0.50 for the same cycle time (T_C = 8 h), lowering the resulting β fraction to 0.34 and theT_{AOX1}/T_E ratio to 0.05 yield $S_{NO} = 6.4 < 7.0$.

Such option (Dual 2), however, is not practicable due to a very short first anoxic period comprising an even shorter filling phase. If a lower T_C of 6 h yielding higher V_0/V_F ratio was selected the same effluent quality could be obtained with higher β and T_{AOX1}/T_E values. Operating at T_C = 8 h, a further decrease in the $S_{NO,\ D}$ limitation to 5.0 mgN/l, requires a three sub-cycle operation *(triple),* with a higher T_{AOX}/T_E ratio of 0.54 divided into T_{AOX1}/T_E = 0.08, T_{AOX2}/T_E = 0.22, T_{AOX3}/T_E = 0.24. Other results are outlined in Table 4.5.

Table 4.5 Selected design parameters and results for step feeding operation complying with three different effluent nitrate requirements

Parameter	Symbol	Unit	Dual-1	Dual-2	Triple
Total anoxic fraction	T_{AOX}/T_E	-	0.40	0.50	0.54
Effective sludge age	θ_{XE}	day	13.3	16	17.4
Nitrified nitrogen	N_{OX}	mgN/l	42	43	43.7
Cycle time	T_C	h	8	8	8
Total sludge age	θ_X	day	17.7	21	23.7
Recycle ratio	V_0/V_F	-	1.2	1.3	1.4
HRT	θ_h	h	17.6	18.4	19.2
The last fill fraction	β	-	0.49	0.34	0.27
Effluent nitrate N	S_{NO}	mgN/l	9.4	6.4	4.9
First anoxic fraction	T_{AOX1}/T_E	-	0.12	0.05	0.08
Second anoxic fraction	T_{AOX2}/T_E	-	0.28	0.45	0.22
Third anoxic fraction	T_{AOX3}/T_E	-	-	-	0.24
First aerobic fraction	T_{A1}/T_E		0.30	0.33	0.17
Second aerobic fraction	T_{A2}/T_E	-	0.30	0.17	0.17
Third aerobic fraction	T_{A3}/T_E	-	-	-	0.12
First fill phase	T_{F1}	min.	42	17	29
Second fill phase	T_{F2}	min.	40	9	29
Third fill phase	T_{F3}	min.	-	-	22

4.2.3. SBR Design for Intermittent Aeration

SBR operation with multiple stages, (*intermittent aeration*), can supply more nitrate for denitrification for a limited recycle ratio and effluent nitrate will probably be determined by denitrification potential. In this type of operation, filling only during anoxic periods to increase the denitrification potential for a given total anoxic fraction may not be practical. Although a part of the denitrification potential is wasted, extended filling throughout the entire process time may still be preferable especially when the S_S fraction of wastewater is relatively low but COD/TKN ratio is high (see Figure 2.3.(c)). Nitrate may not be completely depleted in each sub-cycle due to the frequently shift from mixed to aerated periods. As a result, effluent is determined by the difference between total nitrification capacity, N_{OX} and denitrification potential, N_{DP} of the system.

If the denitrification potential of each anoxic period, $N_{DPi,}$ is sufficient to remove all nitrate introduced from the previous stage (N_{Ai}), oxidized nitrate in the last aerobic period will partly leave the system as effluent nitrate in proportion with V_0/V_F. The effluent S_{NO} concentration can be calculated again from Equation (4.28); in this case, however, β value will be determined by the number of sub-cycles provided that they are of equal duration.

An SBR operation with intermittent aeration and extended filling throughout the entire process time is evaluated as a design exercise for the domestic sewage and the tannery effluent previously tested for pre-denitrification in section 4.2.1, with the main objective to comply with an effluent limitation of $S_{NO,D}$=10mgN/l. For the tannery effluent, the total anoxic fraction, T_{AOX}/T_E and hence the effective sludge age were increased to 0.26 and 13.5d respectively to obtain the required denitrification potential using multiple mixing/aeration. N_{OX} = 112 mg/l and N_{DP} = 102 mg/l values, calculated at this effective sludge age, show that the adopted effluent objective can be achieved provided that N_A is not limiting. As shown in Table 4.6, a selected T_C value of 12 h yields V_0/V_F = 3.2 and a hydraulic retention time, HRT = 50 h. If the resulting effective time, T_E of 10 h is divided into five equal sub-cycles of 2 h, both N_{OX} and N_{DP} will be similarly subdivided and the denitrification potential will be limited in each sub-cycle ($N_{Dpi} < N_{Ai}$). This condition is also satisfied for the first sub-cycle where N_{DP1} = 20.4 mgN (102/5) and N_{A1} = 32 mgN/l as the effluent S_{NO} recycled into the first anoxic zone in proportion with V_0/V_F (10 x 3.2). Consequently, the surplus S_{NO} after each anoxic sub-cycle will accumulate throughout the process to its final effluent level of 10 mgN/l. With this five stage operation mode, the effluent nitrate would be decreased to around 5 mg/l as can be calculated from Equation 4.28, if the operation is so adjusted that denitrification potential is not limiting.

A similar SBR operations with a T_{AOX}/T_E fraction of 0.40 (Domestic-1) and 0.45 (Domestic-2) and T_C of 8 h, six equal sub-cycles of 1 h also provides the N_{DP} limited conditions for the domestic sewage, as outlined in Table 4.6. Although Domestic-2 seems to be under-designed to comply with an effluent limitation of $S_{NO,D}$=10mgN/l, the nitrogen removal performance of this operation will be evaluated with a simulation program in Chapter5.

Table 4.6 Selected design parameters and results for intermittent aeration-extended filling (IAEF) process treating the domestic wastewater and the tannery effluent.

Parameter	Symbol	Unit	Tannery	Domestic-1	Domestic-2
Total anoxic fraction	T_{AOX}/T_E	-	0.26	0.45	0.40
Effective sludge age	θ_{XE}	day	13.5	14.5	13.3
Net heterotrophic yield	Y_{NH}	COD/COD	0.29	0.29	0.30
Nitrified nitrogen	N_{OX}	mgN/l	112	42	42
Denitrification potential	N_{DP}	mgN/l	102	36	31
Effluent nitrate	S_{NO}	mgN/l	10	6	11
Cycle time	T_C	h	12	8	8
Total sludge age	θ_X	day	16	19	18
Recycle ratio	V_0/V_F	-	3.2	1.3	1.2
HRT	θ_h	h	50	18.4	17.6
Oxygen transfer rate	r_{OT}	gO_2/m3.h	53	54	51
Duration of subcycle	t_{SC}	h	2	1	1
Number of subcycle	-	-	5	6	6
Fill time ratio	FTR	-	0.8	0.7	0.7

4.3. DESIGN CONSIDERATIONS FOR PHOSPHORUS REMOVAL

This section describes the basic principles of biological phosphorus removal, factors affecting performance and provides an example for the design of SBR processes that remove only carbon and phosphorus. The processes that are designed to remove both nitrogen and phosphorus will also be dealt with here. These systems, however, have complicated process structure due to many interacting processes involving different microbial consortia acting under different conditions. This makes reliable prediction of effluent quality to be expected from a design almost impossible without the use of simulation programs. The procedures presented here are provided for estimating the required extent of the various periods to use as starting points for simulations. Therefore, this issue will be discussed more detailed in the following chapter.

4.3.1. Mechanism of Biological Phosphorus Removal

It is now widely accepted that *Enhanced Biological Phosphorus Removal*, EBPR, occurs as a result of the predominance of PAOs which are only sustained when the activated sludge system is operated in an anaerobic/aerobic sequence. During the anaerobic period, PAOs can store volatile fatty acids (VFAs) like acetate, S_A, which is mainly produced by ordinary heterotrophs using fermentable substrate, S_F during this period. Storage of VFAs as poly-hydroxyalkanoates, (PHA), proceeds using the internal polyphosphate (poly-P) pool as energy source resulting in a release of orthophosphate (P_i). In the subsequent aerobic period, PAOs grow on the stored PHA and take up P_i for the replenishment of poly- P reserves, thus increasing the P content of the sludge. Because ordinary heterotrophs cannot take up substrates like acetate due to the lack of an electron acceptor, the anaerobic conditions favour the enrichment of PAOs within the activated sludge.

Since sludge wastage is the sole means of removing P in biological treatment systems, the amount of P removed is a function of the net sludge production and the P content of sludge. In entirely aerobic activated sludge systems, P content of the sludge can be assumed constant, as it is determined by nutrient requirements of microorganisms. The P removal in such systems is calculated from the net quantity of sludge wasted which in turn depends on the SRT of the system. In activated sludge systems with an initial anaerobic stage, however, the P content of the sludge is far above the "normal" level (i.e. 0.02gP/gTSS), with the exact value depending mainly upon the fraction of PAOs. Several factors affect the selection of PAOs and hence the overall P content of the sludge required for the desired P removal.

4.3.2. Factors Affecting EBPR

Factors determining the selection of PAOs, and therefore affecting phosphorus removal performance, can be broadly classified into three groups: (i) wastewater characteristics; (ii) system parameters and (iii) environmental factors.

4.3.2.1. Wastewater Characteristics

The concentration of total phosphorus, TP, relative to the biodegradable organic matter in a wastewater significantly affects the performance. The COD/TP ratio of influent is often used as a critical factor to judge whether the process can be operated under phosphorus limited conditions. To sustain low effluent soluble phosphorus (SP) concentrations, phosphorus limited conditions are desired where more than sufficient amount of organic carbon is available to remove all phosphorus present in the influent. Relatively higher SP, however, will be present in the effluent of a carbon limited system at a concentration determined by the difference between influent TP and P removal capability, ΔP. Effluent SP is used for quantification of ΔP because effluent particulate P is mostly associated with suspended solids remaining in the supernatant after settling and therefore not specifically related to biological processes involved. ΔP is determined by the sludge wastage rate and the P content of sludge. The phosphorus content of sludge (P/TSS) remains nearly constant with decreasing influent TP until P becomes limiting:

$$\Delta P = TP_{in} - SP_{eff} = (P_{XT} / Q)P_{cont} \tag{4.31}$$

ΔP, on the other hand, will be almost equal to the influent TP for the phosphorus limited systems and the P content of sludge increases with increasing influent TP until carbon becomes limiting.

The COD fractions of a wastewater significantly affect EBPR performance. Readily biodegradable COD in the form of VFA, S_A, can be used directly by PAOs or becomes available for them after being fermented by facultative heterotrophs under anaerobic conditions. Relatively high anaerobic SRT will be required to generate sufficient VFA when the S_A fraction of the influent COD is low, provided that sufficient fermentable COD, S_F is available in the anaerobic period.

The COD/TKN ratio can also drastically affect the EBPR performance of a nutrient removal SBR system as explained in the previous sections. If the COD/TKN ratio is not high enough to provide the desired nitrogen removal and EBPR for a mixed period fraction increased to the acceptable highest limit, an anaerobic period, and hence EBPR, can only be obtained at the expense of reduced nitrate removal.

4.3.2.2. System Parameters

Sludge retention time and its fractions play an important role in the design of EBPR systems. The aerobic SRT must be selected so as to allow growth of PAOs. A θ_{XA} value of 2 to 3 days can be sufficient enough, depending on temperature (Mamais and Jenkins, 1992). If nitrification is an objective, it will control the selection of the aerobic SRT. If it is not required however, selection of an aerobic SRT value short enough to prevent nitrification may be appropriate since produced nitrate will be recycled to the anaerobic (mixed) period and will interfere with the EBPR process. For simultaneous nitrogen and phosphorus removal systems, care should be taken to adopt the minimum sludge age value that ensures the required nitrification, because longer aerobic (and anoxic) SRTs would lead to a higher total SRT and result in a lower sludge wastage and hence, less P removal. Moreover, excessive aeration causes depletion of stored substrate leading to a *carbon limited* system and secondary release of P due to decay reactions.

The anaerobic SRT required for EBPR is closely related to the influent COD composition. Since storage of VFAs is a very rapid process it can be as short as 0.5 day at 20°C, if the needed amount of VFAs is present in the wastewater (Grady et al., 1999). An anaerobic SRT of at least 1 day is required for the domestic wastewater composition given in the previous sections. The required value increases when hydrolysis of slowly biodegradable substrate is needed before fermentation due to the insufficiency of the readily biodegradable COD in the influent. For simultaneous nitrogen and phosphorus removal systems, the anaerobic SRT must be much longer due to the effects of both increased total SRT and anoxic utilization of readily biodegradable substrate leading to further increase in total SRT. Hence, aerobic and anoxic SRTs beyond the threshold level required to achieve the nitrogen removal objective should be avoided. Pre-fermentation out of the main stream can be considered to increase the VFA content of the influent, if necessary. Effects of system parameters and wastewater characteristics on EBPR performance will be systematically evaluated in the next Chapter.

4.3.2.3. Environmental factors

Decrease of temperature has a lowering effect on the growth rate of PAOs, conveniently expressed by an Arrhenius-type equation, the same way as many similar biological processes. Experimental observations suggest that PAOs are less sensitive to temperature changes as compared to nitrifiers. The proposed default value in ASM2 for the maximum specific growth rate at 20°C is 1.0/day both for PAOs and nitrifiers, which is reduced at10°C to 0.35/day for nitrifiers and only to 0.67/day for PAOs (Henze *et al.,* 1995). This type of a temperature dependency means that the minimum aerobic sludge age for nitrifiers is higher than that of PAOs at temperatures below 20°C and thus controls the selection of the design aerobic SRT in simultaneous nitrogen and phosphorus removal systems. These values also indicate that at temperatures over 20°C it is difficult to prevent concomitant nitrification even when the objective is solely P removal.

The impact of decreasing pH is less significant on PAOs relative to nitrifiers. A similar effect also applies for the dissolved oxygen concentration; sufficient DO concentration must be present in the aerobic period to allow growth of both nitrifiers and PAOs. Due to difference in their half-saturation coefficients for oxygen, a reduction in the DO concentration causes a greater drop in the growth rate of nitrifiers than that of PAOs.

4.3.3 EBPR without Nitrogen Removal

The design of anaerobic/aerobic SBR for only phosphorus and carbon removal relies (i) on the selection of an appropriate aerobic sludge sludge age, θ_{XA}, for the growth of PAOs; (ii) an appropriate anaerobic fraction for sustaining EBPR and (iii) prevention and/or minimization of nitrate intrusion into the anaerobic period by sludge recycle, should nitrification occur.

The aerobic SRT should not be selected, if possible, higher than minimum aerobic sludge age for nitrifiers. Otherwise, nitrification may occur resulting in the presence of nitrate in the effluent and recycle. In this case, higher T_C values resulting in lower recycle ratios, V_0/V_F, should be selected for minimizing a pre-anoxic period. Measures like mixing during idle phase can also be taken (see Figure 2.4.(b)). Deriving accurate analytical expressions for the required anaerobic fraction is almost impossible due to the complexity of processes and the excessive number of components involved. Pilot plant studies or simulation programs implementing appropriate models can be helpful.

The available design methodologies for EBPR systems mostly rely on empirical experience: According to the design procedure proposed for continuous-flow activated sludge systems in the EPA Design Manual (1987), P removal is calculated assuming a constant value of 0.05gP/gTSS for the P content of the sludge, provided that the anaerobic contact time is selected in the range of 0.9-2.0 hours based upon pilot studies. A similar design method based on the concept of the selection of PAOs is given in the WRC Manual (1984) where the P content of the sludge is empirically correlated with the readily biodegradable COD and the anaerobic sludge mass fraction. These procedures may be helpful to initiate a design, however, the only way to predict the P removal performance is to use simulation programs with a calibrated model, such as ASM2d (Henze *et al.,* 1999).

4.3.4. Simultaneous Nitrogen and Phosphorus Removal

SBR processes that remove both nitrogen and phosphorus can be designed using the procedure given in Figure 4.1 with an anaerobic period added to the process sequence. As discussed in Section 2.2.2, SBR operation does not allow elimination of nitrate recycle to the first mixed period when nitrogen removal is desired. However nitrate recycle can be minimized using lower V_0/V_F ratios allowing for anaerobic conditions after a short anoxic phase in the first mixed period. If the wastewater has a high COD/TKN ratio, SBR can provide, as explained in Section 4.1.2, EBPR along with the desired nitrogen removal. This requires long SRTs resulting in relatively low waste sludge production rates. It is therefore problematic to obtain high degrees of nitrogen removal along with EBPR unless the wastewater composition is particularly suitable.

A stepwise procedure for the design of nitrogen and phosphorus removing SBR is illustrated in Figure 4.2. The first six steps of the design are the same as what is defined for nitrogen removing systems in Figure 4.1: After determining θ_{XA}, θ_{XE} and the required recycle ratio for the desired nitrogen removal via predenitrification, the anaerobic fraction of process time, T_{AN}/T_P, is selected according to the desired phosphorus removal. As mentioned above it is difficult to give an analytical expression for the selection of T_{AN}/T_P. It can be increased until $T_M/T_P = 0.6$. If the sum of required anaerobic and anoxic fractions, T_M/T_P, is higher than this value, desired N and P removal cannot be accomplished without external carbon supplement.

A cycle time, T_C, is selected observing the considerations explained in Chapter 3, i.e. HRT and aeration equipment optimization. For domestic wastewaters, the optimum T_C value will likely yield V_0/V_F values lower than the required recycle ratio for predenitrification, R, unless desired denitrification efficiency and/or nitrification capacity is relatively low. Lower recycle ratios should also be preferred to decrease S_S consumption in the pre-anoxic period. In this case, desired denitrification efficiency can be obtained with post denitrification within a *second anoxic* period with no filling. In order to supply more S_S into the anaerobic period, filling in the second mixed period and obviously, in aerobic periods should be avoided. Moreover, filling in the second mixed period may result in secondary release of P_i. The required denitrification potential of the second anoxic period, N_{DP2}, is calculated from the mass balance for nitrate:

$$N_{DP2} = N_{OX} - (1 + V_0/V_F)S_{NO,D} \tag{4.31}$$

Since all S_S will be consumed in the pre-anoxic period and/or stored by PAOs in the anaerobic period, the second denitrification potential can only be generated from endogenous respiration and slowly biodegradable substrate. By virtue of its low utilization rate, substrate stored during the anaerobic period is only partially consumed during the subsequent aerobic period and it is carried over to the second anoxic period, thus enabling a reasonably high denitrification potential. Therefore, T_{AOX2}/T_E to create the necessary denitrification potential can be estimated from the equation below:

$$N_{DP2} = \frac{T_{AOX2}}{T_E}\eta(N_{XS} + N_{ER}) \tag{4.32}$$

The duration of the first aerobic period must be longer than the required time to produce nitrate for post denitrification in order to prevent a secondary anaerobic period. However, it should not be held too long either, to avoid excessive substrate consumption in the aerobic period.

Design exercise for EBPR with nitrogen removal is illustrated using the domestic sewage characteristics (WW1) and the effluent nitrate level of $S_{NO,D}$ = 10 mgN/l specified in section 4.2 to enable comparative evaluation. TP of wastewater is assumed to be 8 mg/l. Nitrogen and carbon mass balances do not change since growth and endogenous respiration are assumed to cease in the anaerobic period. So, the selected and calculated values in the first six steps of the design are as in Table 4.3.

For sustaining EBPR, T_{AN}/T_P is selected as 0.30, a value that should be verified to ensure a total mixed fraction, T_M/T_P, lower than 0.60. Considering the values in Table 4.4, an optimum cycle time, T_C of 6 h is selected with 2 h devoted to settle and draw, leaving 4 h for process phase, T_P. The duration of anaerobic period is calculated as 1.2 h leaving 2.8 h for effective period, T_E that is divided as 1 h for anoxic period(s) and 1.8 h for aerobic period(s), in accordance with the selected T_{AOX}/T_E value of 0.38. Thus, total mixed periods will be 2.2 h yielding a T_M/T_P value of 0.55.

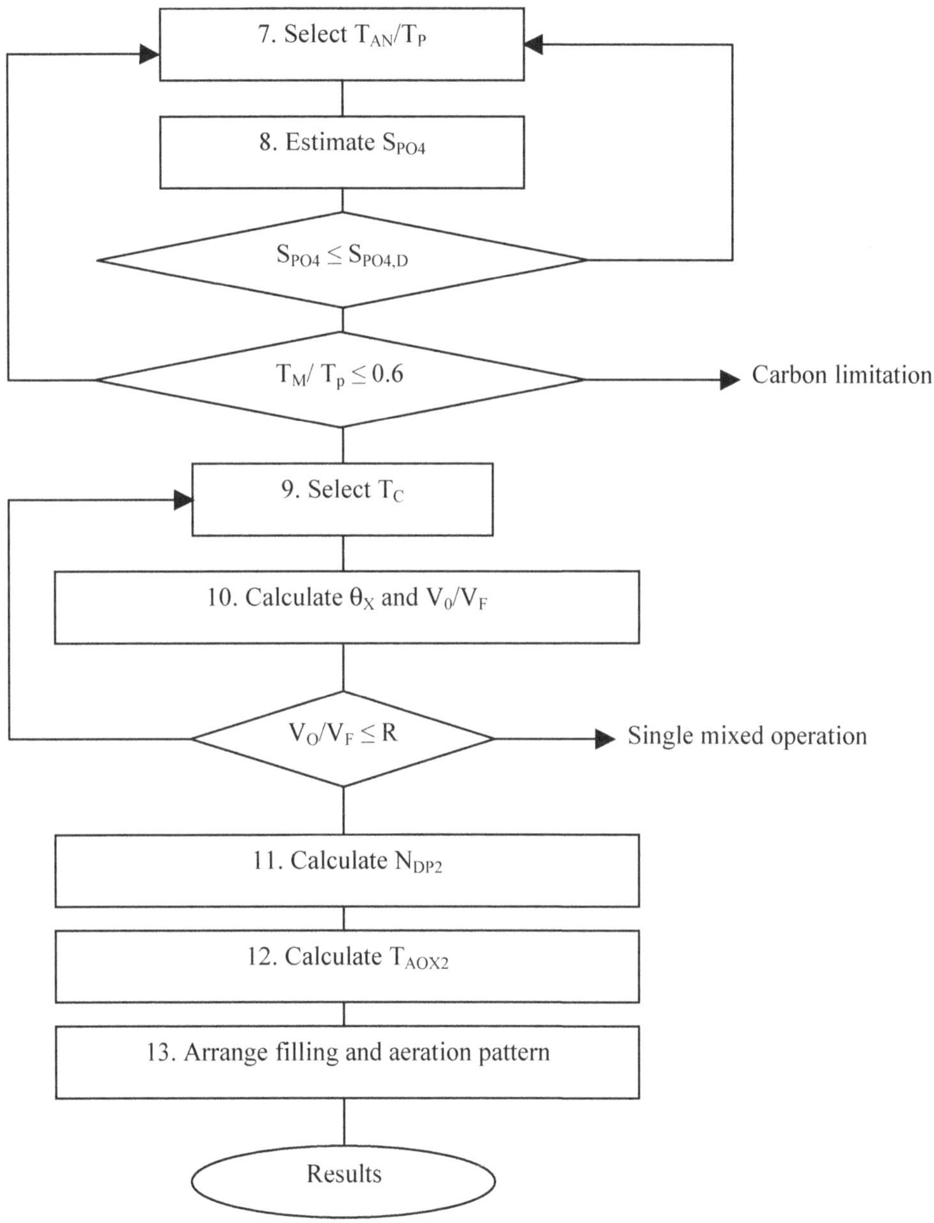

Figure 4.2. The design algorithm for simultaneously nitrogen and phosphorus removal

Addition of an anaerobic period to the process time without changing the effective SRT results in an increase in total SRT and hence V_0/V_F ratio as calculated below:

$$\theta_X = \theta_{XE}\frac{T_C}{T_E} = \theta_{XE}\frac{T_C}{T_A + T_{AOX}} = 12.9(6/2.8) = 27.6 \text{ days}$$

$$\frac{V_0}{V_F} = SF\,(P_{XT}/Q)\,SVI10^{-6}\frac{\theta_X}{T_C} = 1.2*173*100*10^{-6}\frac{27.6*24}{6} = 2.3$$

This value is lower than the required recycle ratio of 3.2 for predenitrification. So, a second stage-denitrification is required to attain the effluent nitrate goal. The required denitrification potential of second anoxic period is calculated as:

$$N_{DP2} = N_{OX} - (1 + V_0/V_F)S_{NO,D} = 42 - (1 + 2.3)10 = 9\text{mgN}/l$$

The required second anoxic time is calculated from Equation 4.32:

$$T_{AOX2} = \frac{N_{DP2}}{\eta(N_{XS} + N_{ER})}T_E = \frac{9}{0.8(50 + 53)}2.8 = 0.3\text{h}$$

The T_{AOX2} period of 0.3 h, subtracted from the total anoxic time of 1.0 h, leaves a first anoxic period of 0.7 h. The fill phase must be shorter than the first mixed period which is 1.9 h, calculated as the sum of first anoxic and anaerobic periods. Thus T_F is selected as 1.5 h yielding a FTR of 1/4. Under these circumstances, approximately 47 % of readily biodegradable COD in the influent will be consumed in the pre-anoxic period (0.7 h/1.5 h). The total aerobic period is divided between the first and second aerobic period as 0.6 h and 1.2 h, respectively. The structure of the cycle is illustrated in Figure 4.3. Selected aerobic and effective sludge ages (8 d and 12.9 d respectively) will secure the desired nitrification and denitrification efficiency as checked in the previous section. The resulting anaerobic SRT, calculated as 5.5 days (27.6x1.2/6 =5.3 d) is expected to be sufficient to remove 8 mg/l P in the influent stream. Design will be verified using a simulation program as shown in Chapter 5.

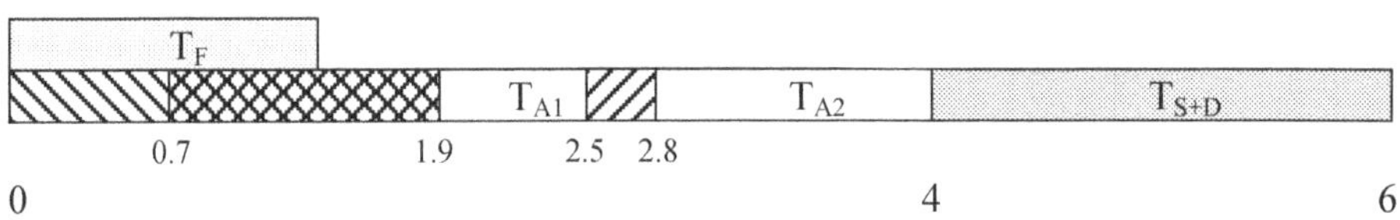

Figure 4.3. Schematic cyclic SBR operation for nitrogen and phosphorus removal from domestic sewage (WW1)

4.4. RETROFIT OF EXISTING SBR PLANTS

Implementation of new effluent limitations involving restrictions for nitrogen and phosphorus concentrations necessitates re-evaluation of the entire stock of existing treatment plants, mainly for retrofitting to nutrient removal. This issue is important for environmentally sensitive areas and especially for coastal areas with small communities and tourist activities.

Existing small-scale SBR systems are generally designed and constructde for organic carbon removal, as a batch version of *extended aeration* activated sludge reactors with a high sludge age (θ_X > 20 days), mainly to minimize excess sludge problems. They are operated with 3-4 cycles per day. The duration of *settle, draw* and *idle* phases (T_S+ T_D+T_I) is usually adopted as 1.5 – 2.0 h. Almost all plants have only one reactor with no additional equalization tank.

Basically, retrofitting should involve an *anoxic* stage, a phase require for the consumption of nitrate nitrogen as electron acceptor and an *anaerobic* stage for enhanced biological phosphorus removal, (EBPR). Phosphorus removal may also be achieved by chemical precipitation. For existing SBR plants, the approach would be to incorporate a *mixed (non-aerated)* phase before – and after if necessary – the aerobic process phase.

The basic idea in retrofitting is to maintain and to utilize with no physical alterations the existing reactor volume. In this context, aside from introducing a non-aerated period into the operation scheme, retrofitting for nutrient removal also requires functional adjustment of the existing total reactor volume, V_T mainly with respect to its two main components, the fill volume,V_F and the initial volume, V_0. This function adjustment should ensure that (i) V_F is adequate to treat the daily wastewater flow for the cycle time, T_C that is appropriate for the selected nutrient removal operation; (ii) V_0 has enough capacity to hold with a safety margin the settled biomass at the end of the cycle, and (iii) the V_0/V_F ratio provides the required level of nitrate recycle into the anoxic period for the targeted effluent S_{NO} discharge.

The retrofitting approach should basically use the same design procedure outlined in previous sections, only modified for a constant θ_h defining the hydraulic residence time of the existing plant. This approach is illustrated below for the wastewater of a typical small community along the Turkish coastal zone with characteristics given in Table 4.7 (Tasli, *et al.,* 2001). The cyclic operation conditions were evaluated to satisfy total nitrogen, C_N = 20mgN/l and total phosphorus, C_P = 3 mgP/l implemented as appropriate effluent limitations. A θ_h value of 16 h based on wastewater flow at full capacity was adopted to characterize the existing SBR plants in the area. The retrofitting exercise was performed for critical summer conditions.

Table 4.7. Seasonal wastewater quality variations for small communities on the southern Turkish coastline (Tasli *et al.*, 2001)

	Small communities*		Domestic Wastewater
Parameter	Off-season	Full-season	Average
COD (mgl^{-1})	480	700	450
TKN(mgl^{-1})	52	50	49
TP(mgl^{-1})	9	10	8
COD/TKN	9.2	15	9.2
COD/TP	53.3	75	56.2
Temp (°C)	16	25	

*hotels, vacation clubs, summerhouses (N<10 000)

Under summer conditions, the aerobic sludge age, θ_{XA} required for full nitrification, may be selected as low as 6 days, due to high wastewater temperature. Then, the high total sludge retention time, θ_X, usually associated with the exsiting SBR reactor designed and operated for carbon removal, may allow for a non-areated period fraction, T_M/T_P, as high as 0.5. This way, the operation of the SBR may conveniently be modified into a nutrient removal system, introducing a single mixed period at the beginning of the cycle. Depending on the balance between N_{DP} and N_A, a significant fraction of the mixed period is likely to sustain anaerobic activities. Similarly, the relatively high COD/N ratio (700/50 = 14) for the wastewater in summer may be regarded as a useful index, inducing a high N_{DP}/N_{OX} ratio and consequently, ensuring effective EBPR. At this stage, consideration of the *filling pattern* becomes important for the calculation of N_{DP}. For a practically convenient continuous filling pattern during T_P, N_{DP}, can be calculated from Equation 4.7 assuming anoxic conditions prevail during the entire mixed period. N_{OX} is calculated from Equation 4.2. For the values selected for sumer conditions in this exercise, the N_{DP}/N_{OX} ratio is computed as 1.2 which indicates that a significant portion of the mixed period will be anaerobic.

When EBPR with partial N removal is envisaged as in this case, this should be achieved with an V_0/V_F ratio as low as possible. A selected value of $V_0/V_F = 1$ secures a denitrification efficiency of 50% yielding an effluent S_{NO} = 17 mgN/l, N_{OX} being 34 mgN/l and consequently N_A will be 17 mgN/l. Then the appropriate cycle time, T_C is computed as 8 h (m = 3) from the basic Equation 2.3, for $V_0/V_F = 1$ and $\theta_h = 16$ h. At this stage, it is posible to compute T_P, which equals $T_C - 2$ h, and T_A as 3 h for $T_M/T_P = 0.5$ and to compute the overall sludge retention time, $\theta_X = 16$ days, from Equation 2.12. This step, where all the parameters are determined for the cyclic operation of the SBR modified for nutrient removal, bring the retrofitting evaluation to a final control mechanism using the fundamental mass balance equation below, which compares the settled sludge volume to be sustained in the reactor for the adopted θ_X and V_0/V_F ratio:

$$\frac{V_0}{V_F} \geq mY_N C_{TI} \theta_X SVI\,10^{-6} \tag{4.32}$$

The numerical evaluation of this expression shows that for the design parameters adopted in this exercise, V_0 remains safely greater than the settled biomass volume provided that the SVI value stays lower than 100 ml/g.

Process stoichiometry also provides a way to evaluate the anaerobic period, T_{AN}. As the available nitrate, N_A controls the denitrification efficiency, T_{AOX}/T_P can be calculated from Equation 4.7 equating N_A to N_{DP}. This equation, numerically solved for this exercise, yields $T_{AOX}/T_P = 0.17$, which leaves an aditional anaerobic period fraction $T_{AN}/T_P = 0.33$. This anaerobic fraction is quite sufficient for the required P removal, since the phosphate storage capacityof the system is a function of the anaerobic biomass fraction, which translates as T_{AN}/T_P for SBR reactors. It should be noted again that process stoichiometry for EBPR is too complex to allow a direct evaluation of the expected removal efficiency; it requires verification by model simulation, which indicates that T_{AN} is long enough for an effective EBPR, reducing the effluent S_{PO4} to 0.02 mgP/l (Tasli *et al.*, 2001).

5

Performance Evaluation by Simulation Models

5.1.NEED FOR SIMULATION MODELS

Appropriate design of a biological plant should ensure desired effluent quality under variable operating conditions. Then, the designer should be able to define, handle and manipulate effective engineering tools for accurate performance evaluation and prediction. A model is needed for this purpose. The model adopted may simply be structured upon personal experience and expertise as it has been common practice in the past. This approach however often fails in the face of unusual conditions, unexpected problems or new requirements. Then, a *physical model* such as a lab-scale reactor or a pilot plant may be used for the purpose of deriving the required information for appropriate design. The physical model, although potentially useful to test and replicate actual plant characteristics, remains site-specific and does not provide alone full understanding of different phenomena controlling treatment processes. This type of an insight can only be acquired and improved by means of *mechanistic models* that would provide a reasonably acceptable description of major biochemical processes in terms of microbial kinetics and stoichiometric relationships among significant parameters. These models, when reliable, may then be used for the identification of potentially feasible engineering solutions.

Mechanistic models have been proposed for activated sludge systems since late sixties (McKinney, 1968; Jenkins and Garrison, 1968). These models, limited to only two components – the overall carbonaceous substrate and biomass, aside dissolved oxygen – and two traditional processes of growth and endogenous decay, remained far from providing an acceptably accurate

 Mechanism and Design of Sequencing Batch Reactors for Nutrient Removal: Scientific and Technical Report No. 19 by Nazik Artan and Derin Orhon. Published by IWA Publishing, London, UK. ISBN: 1843390825.

description of the complex biodegradation processes and relationships among process components. The milestone in the modelling of activated sludge has been the recognition of substrate fractions with different biodegradation rates and the identification of the readily and slowly biodegradable COD fractions (Dold *et al.,* 1980). These concepts together with extensive research efforts on nitrogen removal led to the development of the *Activated Sludge Model No.1,* (ASM1), by Henze *et al.,* (1987) providing a comprehensive description for COD and nitrogen removal. ASM1 incorporates wastewater characterization through COD fractionation including soluble and particulate inert COD components and differentiates active biomass from endogenous residues and other particulate components; it defines heterotrophic growth as a function of the readily biodegradable substrate, a non-energy requiring surface reaction for the hydrolysis of the slowly biodegradable substrate ad a death-regeneration process for the heterotrophic biomass. Orhon and Artan (1994) provided a modified version of ASM1 re-introducing endogenous decay and involving generation of soluble microbial inert COD for a variety of applications including treatment of domestic sewage and and several types of industrial wastewaters. Later, ASM1 was expanded into ASM2 to cover enhanced biological phosphorus removal (Henze *et al.,* 1995). The latter, while providing a satisfactory description for COD and nitrogen transformations, needed improvements for the fate of phosphorus, mainly because it did not rely on sufficient experimental support, and also the behavior of phosphate accumulating organisms (PAOs) did not always appear quite predictable (Kuba *et al.,* 1997; Ekama and Wentzel, 1999). In the light of recent research (Mino *et al.,* 1995; Barker and Dold, 1997; Artan *et al.,* 1998), ASM2 was modified as ASM2d to include denitrification by PAOs with reduced anoxic P uptake (Henze *et al.,* 1999).

As a common feature, mechanistic models describing biochemical processes become more difficult to follow as they improve. Failure to grasp the ever increasing complexity of interactions between system components often leads to serious misinterpretations. In order to overcome this problem, a *matrix format,* composed of *model components* and *processes* constituting the backbone of the model, is usually adopted to reflect the basic stoichiometric relationships. Description of ASM2d in a matrix format is also provided by Henze *et al.* (1999).

Biological nutrient removal involves a wide spectrum of biochemical reactions occurring in an appropriate sequence of aerobic, anoxic and anaerobic conditions. The number of processes and components, together with the complexity of SBR hydraulics, makes it impossible to come up with an accurate prediction of effluent quality without the use of appropriate modelling. The model is first needed to provide an insight on fundamental microbial processes involved. It is only with an acceptably accurate model concept that such processes may be interpreted in terms of relevant mass balances and translated into a design procedure. Model simulation of SBR performance provides information for a selected set of different operating conditions; it is quite useful to visualize the impact of selected operating parameters on systems performance. If performed however, without due consideration of significant microbial mechanisms, it runs the risk of turning into an endless trial and error exercise. Interpretation of the simulation results for process design is only meaningful when support is provided in terms of relevant mass balance relationships between related model components. Aside from a few basic assumptions, effluent quality for carbon removal cannot be evaluated without model simulation. The basic stoichiometry of nitrogen oxidation and removal provides sufficient information for the prediction of system performance, when properly translated into a design procedure for SBR systems, as outlined in the preceding chapter. A similar approach is not currently available for EBPR, due to excessive number of processes and model components involved. Therefore, performance evaluation of SBR systems operated for EBPR still depends upon model simulation. Furthermore, interpretation of the nitrate removal and phosphorus release mechanisms competing for the same organic carbon source in the anoxic/anaerobic phases within simultaneous N/P removal systems also requires model simulation.

5.2. PERFORMANCE EVALUATION FOR COD REMOVAL

Prediction of the effluent COD is important to ensure compliance with discharge regulations. Suspended-growth systems treating domestic sewage are usually over-designed for COD; they provide complete removal of the biodegradable COD in the influent so that the effluent soluble COD consists almost entirely of the soluble inert COD fraction of the treated sewage and the soluble residual metabolic products generated during biological treatment. The magnitude of the total effluent COD also depends on the particulate COD component associated with biomass escaping settling and this way, it is significantly affected by the settling characteristics of the activated sludge.

Performance prediction for COD removal in SBR systems is somewhat more complicated and should take into account (i) different feeding patterns that may be adopted for operation and (ii) COD fractionation of the wastewater: Particulate COD components, X_I and X_S, will be adsorbed onto and contained within the activated sludge flocs. By definition, soluble inert COD, S_I, will not be affected and by-pass the reactor; readily biodegradable COD, S_S, will be totally removed and therefore the magnitude of the soluble COD in the effluent will closely relate to the fate of soluble slowly biodegradable COD, S_H, which constitutes the major portion of the soluble COD in domestic sewage and industrial wastewaters. When the fill time is short at the beginning of the cycle, the SBR system approximates, as previously mentioned, a batch operation and complete hydrolysis and utilization of S_H might be an acceptable assumption, at least for domestic sewage. When the fill time is extended to cover the major portion or the entire cycle for practical convenience, the hydraulic residence time distribution of the feed may not be appropriate to ensure full S_H removal. Simple process stoichiometry and mass balance cannot provide this information and the fate of S_H as a function of the feeding pattern can only be assessed using model simulation.

In this context, the performance of SBR operation is evaluated by means of model simulation for two selected wastewaters with different characteristics, as outlined in Table 5.1: The first wastewater is a typical domestic sewage with a total COD of 525 mg/l and a rapidly hydrolysable COD fraction of 123 mg/l corresponding to 23% of the total COD and 62% of the soluble COD (Orhon *et al.,* 1998). The second one is plain-settled tannery effluent, a much stronger wastewater with a total COD of 2285 mg/l, a higher soluble COD portion and a different distribution of COD fractions (Orhon *et al.,* 1999a).

A simple model, previously tested for a large spectrum of domestic and industrial wastewaters is selected for the evaluation (Orhon and Artan, 1994). The model only involves basic processes and model components necessary for organic carbon utilization. It incorporates endogenous respiration and associated processes for the generation of soluble and particulate residual metabolic products, heterotrophic growth and two-stage hydrolysis. The concept of dual hydrolysis is compatible with COD fractionation, which differentiates soluble and particulate portions of the slowly biodegradable COD. It is introduced by Orhon *et al.,* (1999b) with the argument that it is often misleading to characterize this major COD fraction by means of a single hydrolysis rate, as it is likely to cover a wide array of compounds with different composition and biodegradation characteristics. The matrix representation of the model used in the evaluation is given in Table 5.2. The set of kinetic and stoichiometric coefficients experimentally determined for the selected wastewaters. (Orhon *et al.,* 1999c) is outlined in Table 5.3 and adopted for model simulations.

Table 5.1. COD fractionation of selected wastewaters

Parameter, mg/l	Domestic sewage	Plain-settled Tannery wastewater
Total COD, C_{T1}	525	2285
Total soluble COD, S_{T1}	200	1298
Total particulate COD, X_{T1}	325	987
Total biodegradable COD, C_{S1}	450	1808
Readily biodegradable COD, S_{S1}	52	434
Readily hydrolysable COD, S_{H1}	148	880
Slowly hydrolysable COD, X_{S1}	250	724
Soluble inert COD, S_{I1}	25	64
Particulate inert COD, X_{I1}	50	263

Table 5.2. Endogenous decay model

Processes	S_I	X_E	S_S	S_H	X_S	X_H	S_O	Rate
1. Slow Hydrolysis			1		-1			$k_{hx}\frac{X_S/X_H}{K_{XX}+X_S/X_H}\frac{S_O}{K_O+S_O}X_H$
2. Rapid Hydrolysis			1	-1				$k_{hs}\frac{S_H/X_H}{K_{XS}+S_H/X_H}\frac{S_O}{K_O+S_O}X_H$
3. Growth			$-1/Y_H$			1	$-(1-Y_H)/Y_H$	$\hat{\mu}_H\frac{S_S}{K_S+S_S}\frac{S_{NH}}{K_N+S_{NH}}\frac{S_O}{K_O+S_O}X_H$
4. Decay of Heterotrophs	f_{es}	f_{ex}				-1	$-(1-f_{es}-f_{ex})$	$b_H X_H\frac{S_O}{K_O+S_O}\frac{K_{NO3}}{K_{NO3}+S_{NO3}}$

Table 5.3. Stoichiometric and kinetic coefficients for selected wastewaters

Parameter	Unit	Domestic sewage	Plain-settled Tannery wastewater
Maximum heterotrophic growth rate, μ_{Hmax}	day^{-1}	6.5	2.2
Half saturation coefficient, K_S	$mgCODl^{-1}$	5	28
Maximum hydrolysis rate for S_H, K_{h1}	day^{-1}	3.1	1.1
Half saturation coefficient for S_H, K_{X1}	$mgCOD\ mgCOD^{-1}$	0.2	0.2
Maximum hydrolysis rate for X_S, K_{h2}	day^{-1}	1.2	0.3
Half saturation coefficient for X_S, K_{X2}	$mgCOD\ mgCOD^{-1}$	0.5	0.2
Endogenous decay coefficient, b_H	day^{-1}	0.15	0.12
Heterotrophic yield, Y_H	$mgcellCODmgCOD^{-1}$	0.64	0.64
Particulate inert fraction of endogenous decay, f_{EX}	-	0.20	0.20
Soluble inert fraction of endogenous decay, f_{ES}	-	0.07	0.07

5.2.1. Effect of Filling on Effluent Soluble COD

In this section, the effect of the adopted wastewater filling pattern on the degree of S_H utilization and therefore on the resulting level of effluent COD is evaluated by model simulation. For domestic sewage, the SBR operation selected for model evaluation includes typical parameters, i.e. an effective sludge age of 10 days, a total cycle time, T_C of 6 hours, a fully aerated effective period, T_E of 4 hours and a hydraulic residence time of 12 h with $V_0/V_F = 1.0$. Three different filling patterns of fast fill (1 hour), slow fill (3 hours) and extended fill (4 hours) covering the entire effective cycle time are tested in process performance simulations. The results obtained are given in Figure 5.1, which illustrates concentration profiles of soluble and particulate COD fractions in the course of the operating cycle.

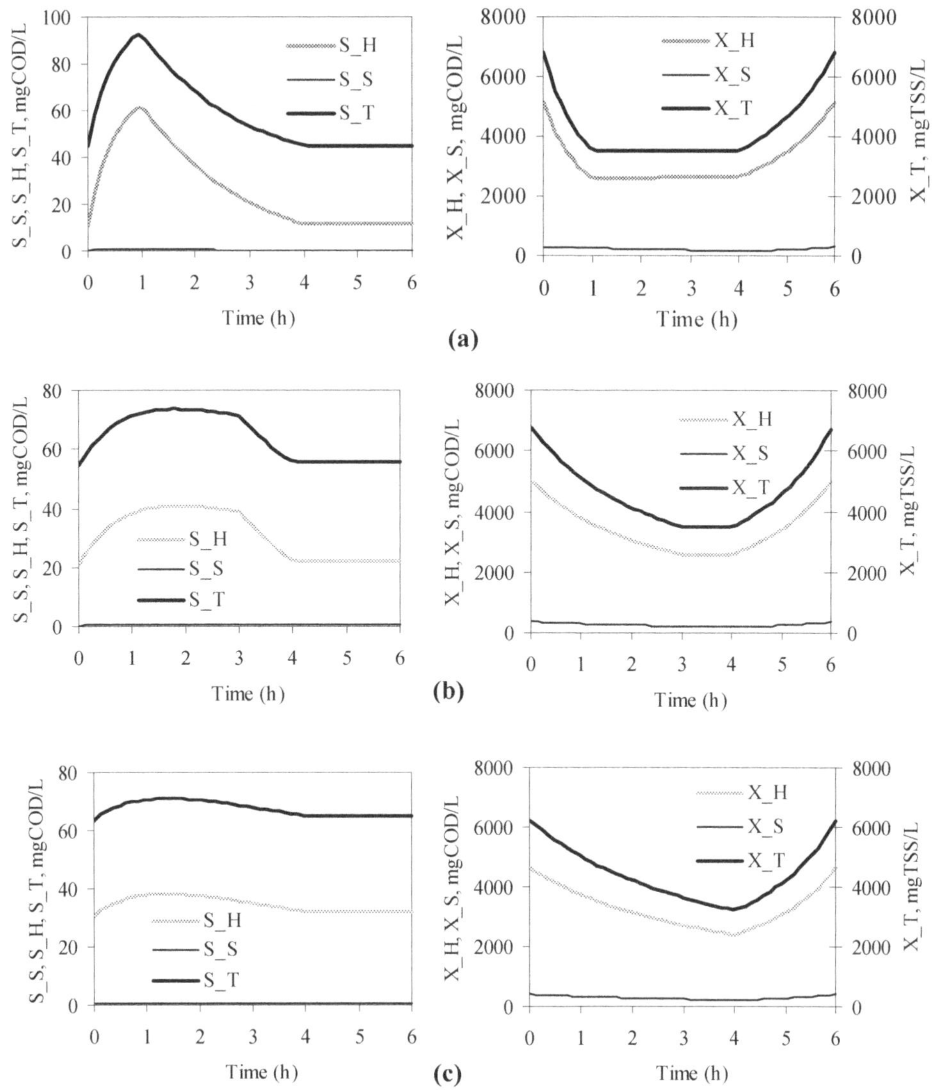

Figure 5.1 COD removal performance for domestic wastewater. (a) Fast Fill (b) Slow Fill (c) Extended Fill

Similarly, evaluation of tannery wastewater treatment was made for an SBR operation for a higher θ_{XE} level of 16 days, with selected T_C and T_E values 12 and 10 h respectively result in $V_0/V_F = 1.5$. The fill time ratio, FTR was set as 2/12 for the fast fill mode, 6/12 for the slow fill mode; extended fill continued for the whole effective cycle time (FTR = 10/12), as for domestic sewage. COD profiles resulting from model simulation are shown in Figure 5.2.

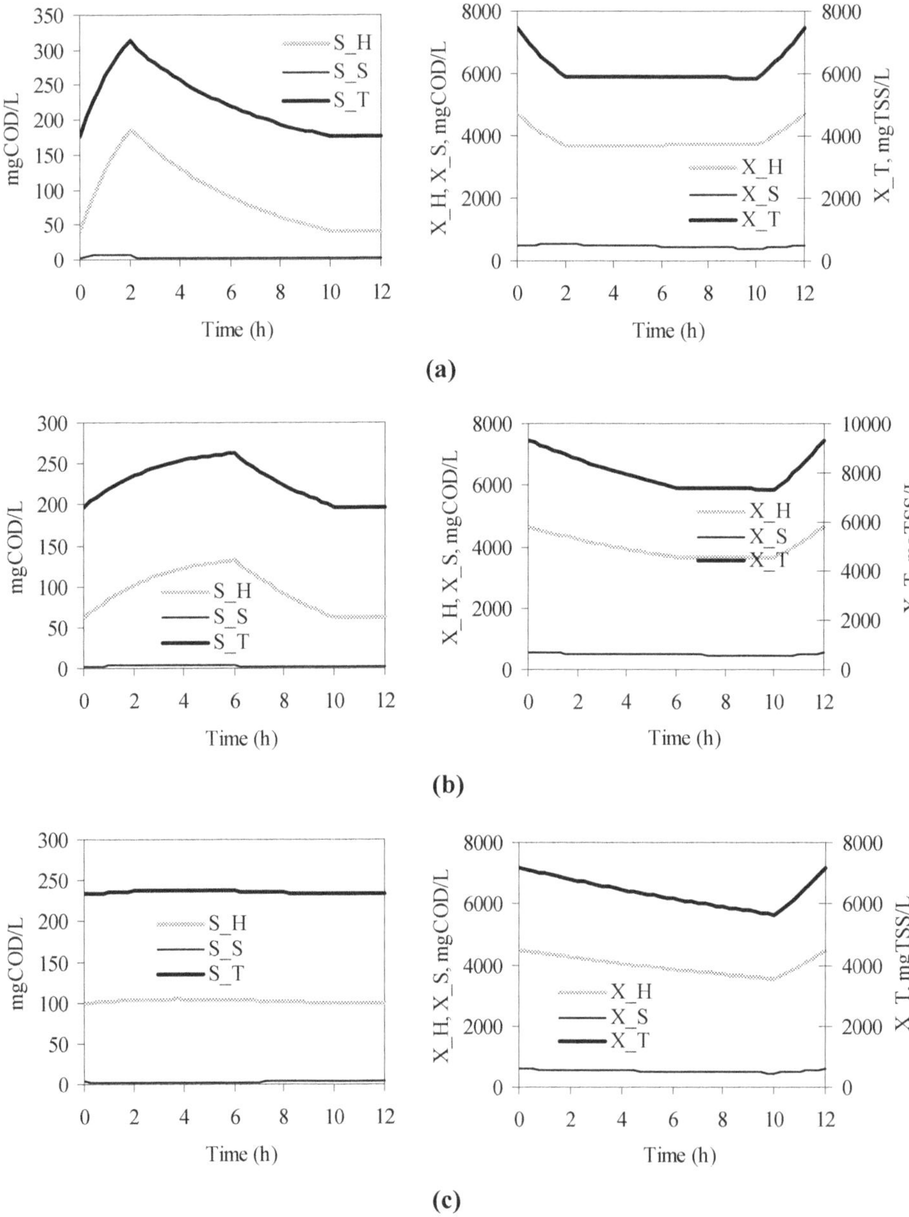

Figure 5.2 COD removal performance for tannery wastewater. (a) Fast Fill (b) Slow Fill (c) Extended Fill.

The results obtained indicate that the fill time ratio has an appreciable effect on the effluent total soluble COD, (S_T), levels both for domestic sewage and tannery wastewater. This effect is better visualized in Figure 5.3 showing the variation of the S_T profiles within the cyclic operation as a function of the three selected FTR values: Increased FTR is observed to induce a change in the effluent S_T from 45 mgCOD/l to around 70 mg/l for domestic sewage. Similarly, a change in the fill mode from fast fill to extended fill exerted a more pronounced S_T increase in the range of 180 – 240 mgCOD/l for tannery wastewater.

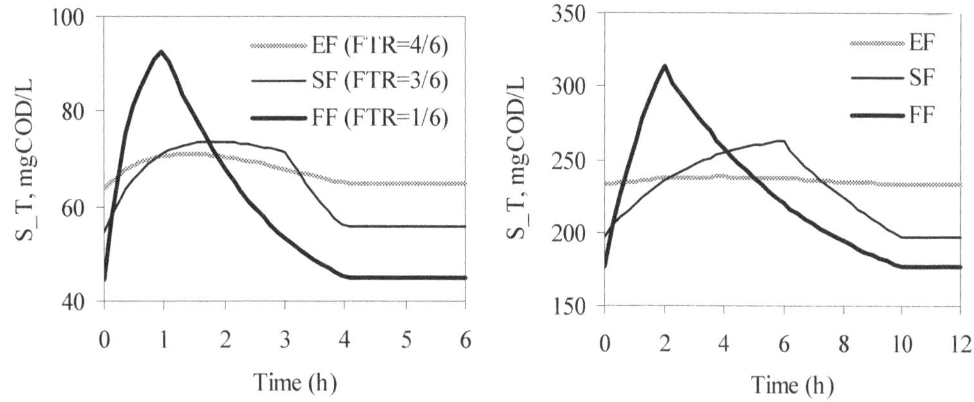

Figure 5.3. Effect of Fill Time Ratio on Effluent Soluble COD (a) Domestic sewage; (b) Tannery Wastewater.

5.2.2. Effect of Filling on Dissolved Oxygen Profile

The effect of three different schemes previously selected for domestic sewage is also evaluated for the resulting dissolved oxygen, (DO) profiles during the cycle, using the same operating parameters. The oxygen transfer coefficient, K_La associated with the aeration capacity provided for the system defines in principle a constant, average DO level which is only indirectly relevant, since in reality a changing DO profile is established due to the transient nature of the SBR operation. The fill mode of the operation is bound to have a significant impact on the DO profile as it determines the pace of the biochemical reactions within the cycle. The evaluation is made for two different K_La values of 110 and 160/day, the first one corresponding to an average DO concentration of around 2.0 mg/l for selected wastewater character and operating conditions. The evaluation results given in Figure 5.4 show an increasing DO trend for both K_La values, markedly different for fast fill as compared to other filling patterns, reaching at the end of the process phase, a DO value increasing in the range of 1.5 – 3.0 mg/l for K_La = 110 d^{-1} and 4.0 – 5.0 mg/l for K_La = 160 d^{-1} as the adopted fill time ratio is lowered. The DO profile within the cycle may have significant effect on both process performance and population dynamics and the effect of FTR should also be considered in this respect.

Furthermore, it is interesting to note that no direct analogy is allowable between an SBR operation with extended filling and a continuous-flow complete mixed activated sludge reactor, simply because of significant transients in DO, despite somewhat constant S_T profiles throughout the cycle.

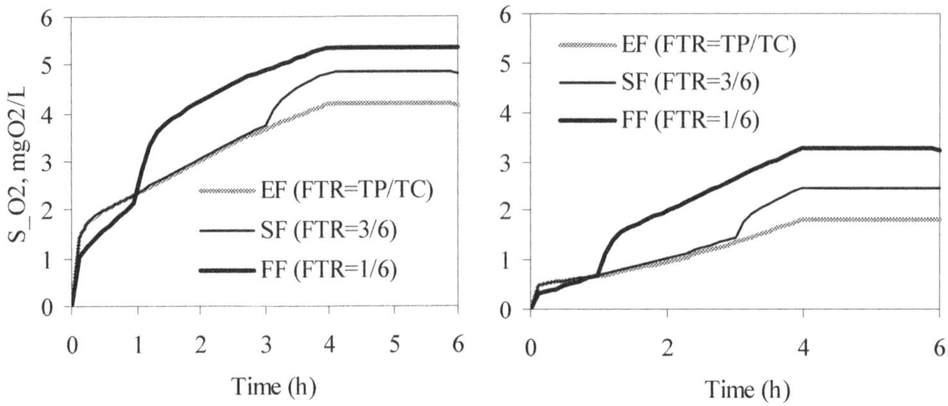

Figure 5.4. Effect of Fill Time Ratio on the Dissolved Oxygen Profile for Domestic Sewage; (a) K_La= 110 d^{-1}; (b) K_La = 160 d^{-1}.

5.3. PERFORMANCE EVALUATION FOR NITROGEN REMOVAL

In this section, nitrogen removal performances of various processes are evaluated using a simulation program, AQUASIM (Reichert, 1998), that implements simplified ASM2 without the EBPR module. Provision of a detailed description of ASM2 is beyond the scope of this report and can be found elsewhere (Henze *et al.,* 1995). Basically, this dynamic model includes hydrolysis, growth of facultative heterotrophic and autotrophic organisms and the decay processes for both organisms modelled with the death–regeneration concept. This process is first order with respect to the active biomass concentration, however, the rate coefficient is different, conceptually and therefore in magnitude, from the endogenous respiration coefficient used in the steady-state design model. Contrary to the endogenous respiration approach, no loss of COD is involved in this process. The decay coefficient, b'_H, in the model must be larger to give the same amount of electron acceptor utilization per time due to decay. It is calculated as 0.35 d^{-1} from equation below using the Y_H and b_H values adopted in the design examples and assuming f_P = 0.1 for the model coefficient indicating the fraction of inert endogenous biomass after loss of viability:

$$b_H' = \frac{b_H}{1 - Y_H(1 - f_P)} \quad (5.1)$$

The other necessary kinetic and stoichiometric coefficients, suggested as default values at 20 ^{0}C in ASM2 were selected for the simulation runs. Simulations were carried out first with inputs already obtained using the design procedure applied for both the domestic and the tannery wastewaters given in section 4.2 to compare the modelling outputs with the design obtained from basic mass balance and process stoichiometry. The factors leading to the difference between calculated and simulated effluent concentrations of nitrogen species will be explained. Some aspects of design requiring simulation models, like K_La's effect on process performance, will also be dealt with here.

5.3.1. Evaluation of Simulation Results

The model simulation basically involved five runs for domestic sewage and four runs for tannery wastewater, with different process alternatives and operating conditions. For domestic sewage, Run 1 (D 1) simulated *pre-denitrification*, Runs 2 and 3 (D2; D3) *step feeding* with dual and triple anoxic fill phases and Runs 4 and 5 (D4; D5) *intermitted aeration* involving extended filling along the entire process time. Simulation of tannery wastewater treatment only involved *pre-denitrification* (T1;T2) and *intermittent aeration* (T3:T4) process configurations. The simulation may duplicate two different operation options with regard to aeration: the DO concentration during the aerobic periods may be set to a specific value, as can be sustained by a DO controller which would adjust the value of mass transfer coefficient, K_La of the aeration equipment. Alternately, aeration for the aerobic periods can be defined in terms of a given K_La value resulting in a DO profile imposed by the selected operating conditions, as explained in the previous section. For the first four Runs of domestic sewage simulation was carried out with a constant DO concentration of 4 mg/l, which was reduced to 2.0 mg/l in Run 5, allowing this way to visualize the effect of DO on the performance of intermittent aeration. Conversely, a set K_La value of 300 d^{-1} was adopted in the Runs designed for tannery wastewater; Run 2 (T2) was performed with a lower K_La value of 140 d^{-1} providing a second evaluation for the pre-denitrification mode of operation. In Run 4 (T4), the aerobic sludge age of 10 days was lowered to 8.6 days by increased anoxic fraction without changing total SRT, for the evaluation of IAEF process under available nitrate limited conditions.

Table 5.4 shows important design and operation parameters for the simulation runs for domestic sewage and effluent values of simulated nitrogen species at steady state together with the effluent S_{NO} calculated from the design procedure based on mass balances elaborated in section 4.2, to allow comparative evaluation. Operation at steady state was obtained by relaxation over several SRTs. The corresponding S_{NH} and S_{NO} profiles associated with the cyclic operation of the SBR systems are illustrated in Figure 5.5. Similar data related to tannery wastewater are displayed in Table 5.5 and illustrated in Figure 5.6.

Table 5.4 Simulation of nitrogen removal performances of various processes designed for nutrient removal from domestic sewage.

RUN	Process	T_{AOX}/T_E	T_C (h)	SRT (d)	V_0/V_F -	HRT (h)	T_A/T_C	Calc. S_{NO} mg/l	Sim. S_{NO} mg/l	Sim. S_{NH} mg/l
D1	PD	0.38	4	25.8	3.3	17.2	0.31	9.8	9.6	0.2
D2	SF-dual	0.40	8	17.7	1.2	17.6	0.45	9.4	9.8	0.2
D3	SF-triple	0.54	8	23.7	1.4	19.2	0.34	5.0	5.0	0.6
D4	IAEF	0.40	8	17.7	1.2	17.6	0.45	11.0	9.7	0.5
D5	IAEF*	0.40	8	17.7	1.2	17.6	0.45	11.0	6.8	0.7

* DO=2 mg/l

Results generally reflect a reasonable agreement of the simulated effluent concentrations with the calculated values, especially for systems where performance is not controlled by N_{DP} limitation. Possible factors causing the difference between the simulated and calculated effluent values can be mainly classified into three groups, namely (i) N_{DP} prediction; (ii) aeration intensity and (iii) effluent S_{NH}. Simulation programs are most useful in improving design based on mass balances where the effect of these factors is estimated to be significant.

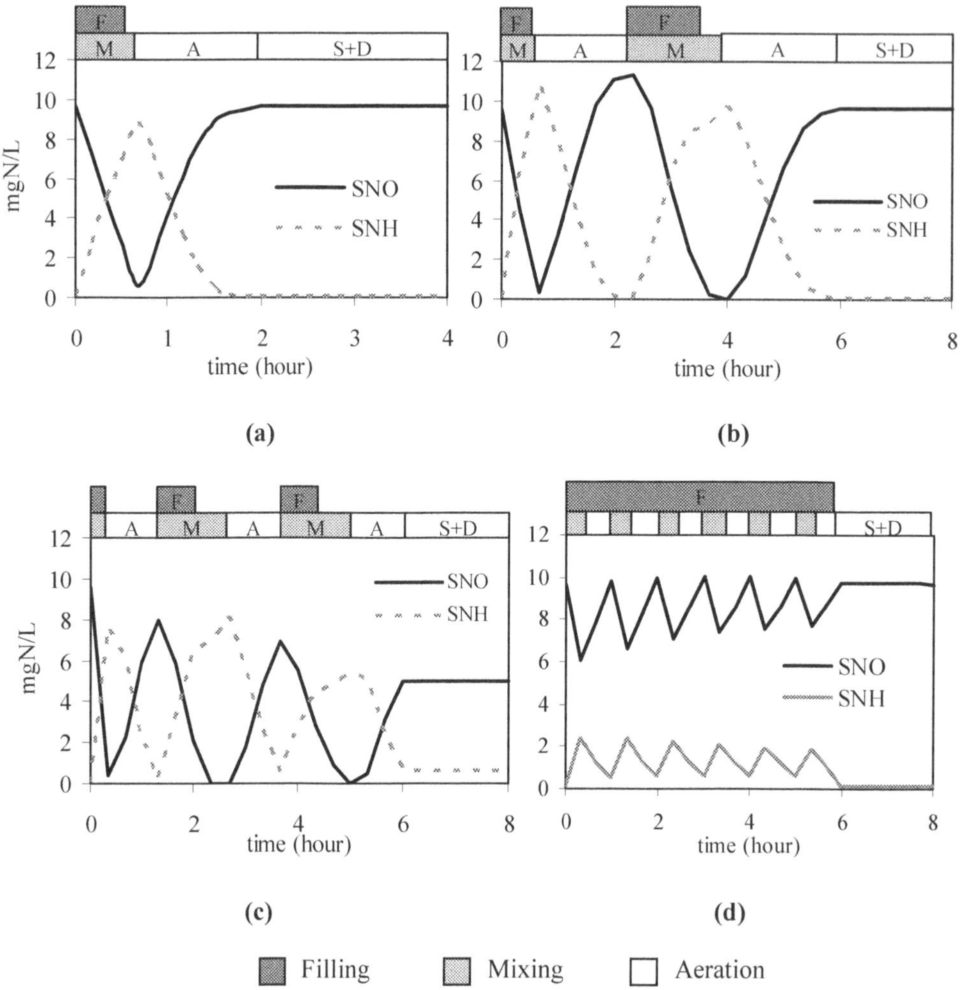

Figure 5.5 Nitrogen profiles for domestic wastewater. (a) Run D1 (b) Run D2 (c) Run D3 (d) Run D4

Table 5.5 Simulation of nitrogen removal performances of various processes designed for tannery wastewater.

RUN	Process	T_{AOX}/T_E	T_C	SRT	V_0/V_F	HRT	T_A/T_C	Calc. S_{NO}	Sim. S_{NO}	Sim. S_{NH}
		-	(h)	(d)	-	(h)	-	mg/l	mg/l	mg/l
T1	PD	0.10	8	15	4.9	47	0.66	16.3	14.4	0.4
T2	PD*	0.10	8	15	4.9	47	0.66	16.3	11.0	0.6
T3	IAEF	0.26	12	16	3.2	50	0.62	10.0	4.5	1.2
T4	IAEF	0.35	12	16	3.2	50	0.54	5.3	4.0	1.5

* $K_La=140\ d^{-1} < K_La$ for the other runs

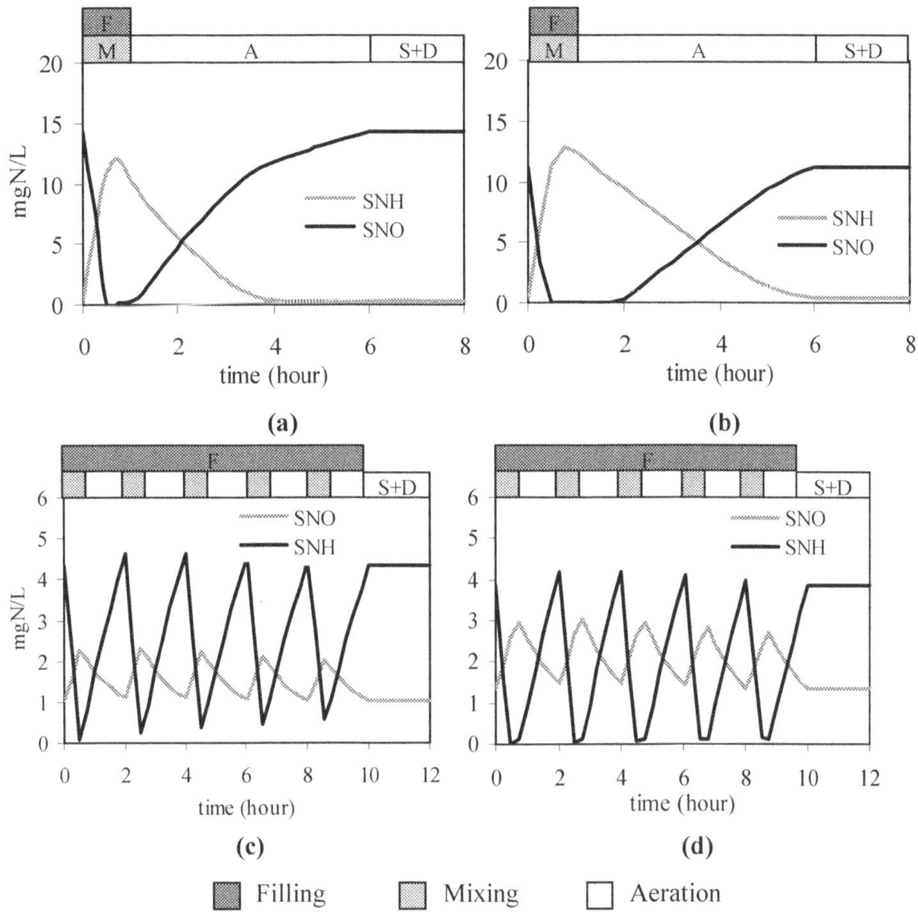

Figure 5.6 Nitrogen profiles for tannery wastewater. (a) Run T1 (b) Run T2 (c) Run T3 (d) Run T4

5.3.1.1. Effect of N_{DP} Prediction on Effluent Nitrate Nitrogen

For predenitrification systems, effluent nitrate, S_{NO} is determined by the recycle ratio and nitrification capacity provided that the denitrification potential of the anoxic period is sufficient. In Figure 5.5 for example, the simulated nitrate profile for Run D1 shows that recycled nitrate (R S_{NO} =31 mgN/l) is almost depleted at the end of the anoxic period. DO intrusion into the mixed period is assumed to be negligible. This implies that denitrification potential of the system is about 31 mg N/l slightly lower than the calculated one given in Table 4.3. In Run T1, operation at high K_La of 300 d^{-1} results in an equally high DO concentration of 6.7 mg/l at the end of the aerobic period, which is recycled back to the mixed period. As shown in the Figure 5.6 (a), nitrate drops to zero just before the aerated period starts, leading to conclude that the denitrification potential of the system is high enough to utilize the total electron acceptor introduced into the mixed period:

$$(S_{NO} + \frac{S_{O2}}{2.86})\, V_0 / V_F = (14.5 + \frac{6.7}{2.86})4.9 = 83 \text{mgN/l}$$

This value is comparable with the calculated N_{DP} value given in Table 4.3. Therefore, N_{DP} prediction for predenitrification systems used in the design model seems to be fairly reasonable.

If available nitrate is not limiting as in Run D4 and T3, N_{DP} directly controls the nitrate concentration in the effluent and hence, accurate N_{DP} prediction becomes decisively important. Simulation results show that N_{DP} prediction according to the selected anoxic fraction using Equation 4.7, underestimates the real situation for systems where simultaneous nitrification/denitrification are likely to take place, as further elaborated in the next section. In such cases, simulation programs should be used to improve the design, where appropriate wastewater characterization is available.

5.3.1.2. Effect of Aeration Intensity

Denitrification potential of a system is affected not only by the nominal anoxic fraction but also by the DO concentration profile in the aerobic period(s). A positive contribution should be expected from denitrification likely to take place together with nitrification during the aerated periods. The conventional physical explanation for this phenomenon is that simultaneous nitrification and denitrification may occur as a consequence of dissolved oxygen (DO) concentration gradients within microbial flocs, due to diffusion limitations. Simultaneous denitrification (SDN) may proceed at different rates dictated by the selected values for DO concentration and the oxygen half saturation coefficient for the heterotrophs, $K_{O,H}$, according to the switch function of $K_{O,H} / (K_{O,H} + S_{O2})$. For the DO level set to 4.0 mg/l in runs D1 to D4 and a default value of 0.2 mg/l for $K_{O,H}$, model predicts "an aerobic period denitrification rate" corresponding to 4.8% of what is normally achieved under anoxic conditions, while with the DO level of 2.0 mg/l in Run D5, this ratio increases to 9.1%. Comparing runs D4 and D5, it is observed that lower average DO concentrations in the aerated periods lead to significant SDN resulting in a substantial S_{NO} difference with respect to design results based on basic mass balance.

The specified DO concentration cannot be attained for a while in the beginning of the aerated period(s) at a given constant K_La. For predenitrification processes that have relatively low FTR, readily biodegradable substrate concentration may still be high in the transient period between anoxic and aerobic periods; this may result in considerable SDN due to the relatively high denitrification rates. Consequently, the simulated effluent nitrate will be lower than the value calculated from mass balance given in Equation 4.13 for the predenitrification process. This difference is clearly observed in Run T2 in which a lower K_La was used as compared to Run T1. The SDN may also be effective in the case of IAEF mode; transient periods last longer due to intermittent aeration and readily biodegradable substrate concentration during the process phase may be relatively high due to the extended filling. So, the real denitrification potential will be higher than the calculated value according to the nominal anoxic fraction leading to lower simulated effluent nitrate on the condition that SDN is higher than the wasted denitrification potential due to oxygen introduced into the anoxic periods. Tables show that simulated effluent S_{NO} concentrations for IAEF process are fairly lower than the calculated ones. This difference is more observable in the Run T3 where S_{NO} is determined by the denitrification potential. In the Run T4, however, a high anoxic fraction ensures sufficient denitrification potential, and a part of the denitrification potential is lost during the previous sub-cycles, thus the effluent nitrate is determined by the available nitrate. Therefore, the difference between calculated and simulated values is presumably due to SDN proceeding only in the last aerobic period.

5.3.1.3. Effects on Effluent Ammonium Nitrogen

In the design procedure, complete nitrification (effluent $S_{NH} \leq 1$ mg/l) is assumed to be achieved for the aerobic SRT, which is selected considering maximum specific growth rate for autotrophs at a design temperature and the average DO concentration in the aerobic period(s). The DO concentration affects nitrification rate by a factor of S_{O2} / ($K_{O,A}$ + S_{O2}) which is calculated, for example, as 0.8 for S_{O2}=2mg/l and a default $K_{O,A}$=0.5 mg/l. The DO concentration profile during the aerated periods cannot be accounted for without the use of dynamic models. As illustrated in the previous section, DO changes during the aerated period for a constant K_La, and FTR exert a significant effect on the DO profile. The filling pattern also may have an important effect on soluble substrate concentration (S_{NH} in this case) during the process time as discussed in section 5.2. The simulation program computes the effluent ammonia nitrogen using nitrification rates limited by both S_{NH} and S_{O2} profiles. So, the exact value of effluent S_{NH} can only be computed with simulation programs which utilize dynamic models.

Filling and aeration pattern is observed to induce an effect on the effluent S_{NH}, as can be seen from Table 5.4 and 5.5. Comparisons of Runs D1 to D4 and T1 to T3 show that a higher effluent value is obtained with the extended filling mode. The reduced K_La and reduced aerobic SRT also affect the effluent S_{NH}: The operation mode illustrated by Run T4 where the aerobic SRT was reduced from 10 d to 8.6 d by increased anoxic fraction at constant total SRT, resulted in a slightly increased effluent S_{NH}.

The effect of K_La on effluent S_{NH} is demonstrated with the Run T2 where K_La is reduced from 300 d^{-1} to 140 d^{-1} for a constant aerobic SRT. A decrease in aeration intensity inevitably causes lower DO concentrations and consequently higher effluent S_{NH} levels. As shown in Table 5.5, the effluent S_{NH} increased slightly for the K_La value of 140 d^{-1}; Model simulation would show a further decrease of K_La down to 100 d^{-1}, would result in total elimination of nitrification at this temperature.

5.3.2. Comparison of Process Options

Table 5.4 shows that the same nitrogen removal performances can be obtained with approximately the same reactor volumes but different operation conditions. The other factors like investment and running costs of aeration equipment, equalization volume requirements and operational simplicity play important role in the process selection (see Chapter 3). The pre-denitrification process (PD), is observed to yield the best effluent quality with the lowest HRT when DO concentration is kept constant at 4mg/l in the aerated periods. However, lower aerated fraction, T_A/T_C, likely to result in higher aeration equipment cost and lower FTR required for predenitrification will cause operational complexity and/or higher holding tank volume compared to the IAEF process.

Simulation results also indicate the importance of DO control especially for the IAEF process, not only in terms of operational cost but also for achievable concentrations of nitrogen species in the effluent. For instance, effluent nitrate would range from 5 mgN/l to 16 mgN/l without DO control in this process depending on the selected K_La ranging from 95 d^{-1} to 140 d^{-1} with no significant effect on the effluent S_{NH}. Comparing these values with the K_La value of 200 d^{-1} required to sustain nitrification for the predenitrification process confirms that larger aeration equipment requires for the same performance in the predenitrification mode.

5.4. PERFORMANCE EVALUATION FOR PHOSPHORUS REMOVAL

Enhanced biological phosphorus removal, (EBPR) involves a sequence of perhaps the most intricate processes utilized for biological wastewater treatment. The complexity of related biochemical processes does not allow a relatively simple evaluation based on stoichiometry, which would lead to basic mass balance expressions directly incorporated into a rational design approach. Instead, model simulation of SBR performance for EBPR may provide useful information for a selected set of different operating conditions. Model simulation is most useful when implemented to visualize the effect of significant factors on system behaviour and process performance.

In this section ASM2 implemented by AQUASIM was used for the simulation of phosphorus removal in SBR system. The simulation was carried out for domestic sewage, mainly to emphasize the effect of significant design and operation parameters, together with wastewater composition, on SBR performance. Simulations were performed for an SBR with a sequence of only anaerobic-aerobic periods within the process phase, with the assumption that no nitrification takes place, excluding this way possible effect of nitrate utilization on phosphorus removal. The effect of nitrate will be covered in the next section.

5.4.1 Effect of System Parameters

To provide continuity, the same domestic wastewater composition used for the simulations in the previous section, with a total COD of 525 mg/l, a total biodegradable COD of 450 mg/l and a readily biodegradable COD of 52 mg/l, was adopted for assessing the P removal capability of the SBR system operated with selected parameters. The readily biodegradable COD was assumed to include a fermented, (S_A), fraction of 20 mg/l and a fermentable, (S_F), fraction of 32 mg/l respectively. The defaults values suggested in ASM2 for the kinetic and stoichiometric coefficients associated with PAOs at 20 ^{0}C, were utilized for the simulation runs. A relatively high influent total phosphorus, (TP), concentration of 15 mgP/l, which would not allow total P removal and thus, a better analysis of the effect of selected system parameters on SBR performance, was selected for the evaluation. The basic design of the SBR system required to run the simulation program was performed in accordance with the design procedure previously defined for organic carbon removal, involving a constant cycle time, T_C of 6 h, a process time, T_P of 4 h and a range of total sludge retention times between 5 to 20 days. For each operation with a selected total SRT, the anaerobic fraction of the process time, T_{AN}/T_P was changed within the range of 0.20 – 0.30. It should be noted that the remaining portion of T_P corresponds only to the aerobic period, T_A, in these simulations since nitrification process is assumedly inhibited to eliminate the effect of recycled nitrate. Fill phase was assumed to continue throughout the anaerobic period ($T_F = T_{AN}$). The selected operation conditions and simulation results are outlined in Table 5.6.

The results of performance simulations may be interpreted in a number of different ways, all quite useful for an accurate evaluation of SBR behaviour under different operating conditions. It first indicates that the effect of the total SRT on phosphorus removal, ΔP, can best be visualized together with the T_{AN}/T_P selected for operation. At this stage, it should be mentioned that a process sludge age, θ_{XP} has to be defined for EBPR systems differently from the effective sludge age, in a way to cover the entire process phase, T_P:

$$\theta_{XP} = \theta_X \frac{T_P}{T_C} \tag{5.2}$$

Table 5.6 Effect of system parameters on Phosphorus removal for domestic wastewater.

θ_X (d)	θ_{XP} (d)	T_{AN}/T_P -	T_{AN} (h)	T_A (h)	θ_{XAN} (d)	θ_{XA} (d)	V_0/V_F -	S_{PO4} mg/l	ΔP mg/l	P_{cont} %
5	3.33	0.20	0.8	3.2	0.7	2.7	0.6	10.5	4.5	2.0
5	3.33	0.25	1.0	3.0	0.8	2.5	0.6	10.5	4.5	2.0
5	3.33	0.30	1.2	2.8	1.0	2.3	0.6	6.4	8.6	3.7
5	3.33	0.50	2.0	2.0	1.66	1.66	0.6	10.3	4.7	2.2
10	6.66	0.20	0.8	3.2	1.3	5.3	1.1	9.6	5.4	3.1
10	6.66	0.25	1.0	3.0	1.7	5.0	1.1	4.6	10.4	5.1
10	6.66	0.30	1.2	2.8	2.0	4.7	1.1	0.7	14.3	6.3
20	13.3	0.20	0.8	3.2	2.7	10.7	1.3	8.3	6.7	4.3
20	13.3	0.25	1.0	3.0	3.3	10.0	1.3	5.3	9.7	5.6
20	13.3	0.30	1.2	2.8	4.0	9.3	1.3	4.0	11.0	6.4

As shown in Figure 5.7, a T_{AN}/T_P ratio of 0.20 is too low to initiate and sustain appreciable EBPR and increasing the total SRT does not provide a significant improvement. At higher T_{AN}/T_P ratios a higher total SRT results in ΔP increase up to θ_X = 10 days but a further θ_X increase deteriorates the P removal efficiency.. For a given process SRT, θ_{XP}, the anaerobic sludge age, θ_{XAN} and consequently ΔP increases with increasing anaerobic fraction as shown in Figure 5.8, however, the ΔP increment is slower at high SRTs since aerobic SRTs also increase together with anaerobic SRTs. It should be noted that when θ_{XP} is low, there is no EBPR at low T_{AN}/T_P ratios, simply because of insufficient fermentation due to lower θ_{XAN}; ΔP increases with increasing T_{AN}/T_P ratios within the range of 0.20 – 0.30. However further increase in anaerobic fraction (for example 0.50) causes a θ_{XA} below the washed-out value of PAOs and ΔP decreases again as given in Table 5.6.

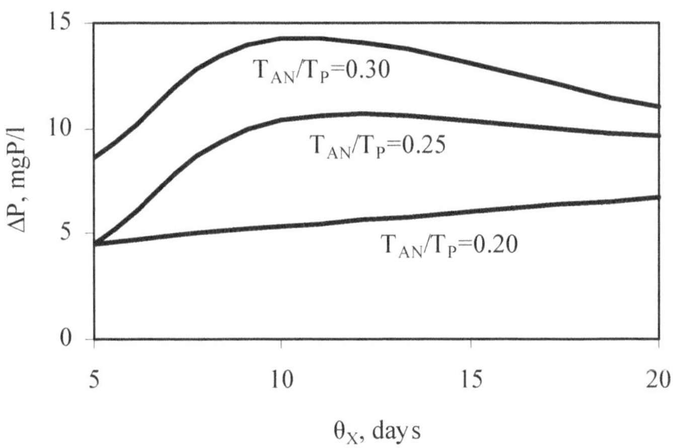

Figure 5.7. Effect of θ_X on phosphorus removal for various anaerobic fraction

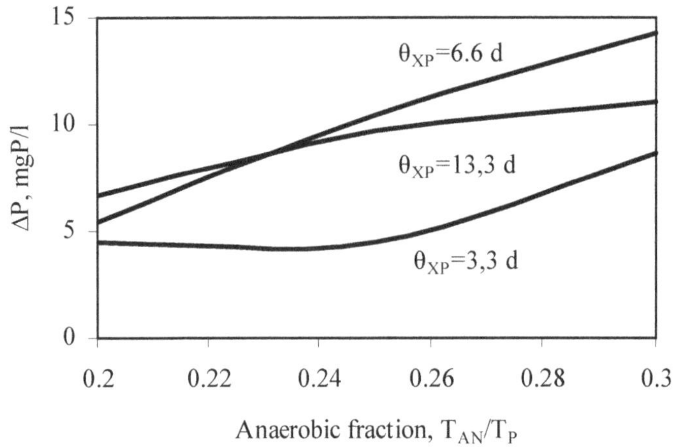

Figure 5.8. Effect of anaerobic fraction on phosphorus removal for various θ_{XP}

Figure 5.9 shows the effect of anaerobic sludge age, θ_{XAN} on phosphorus removal for the two selected aerobic sludge age values of 3 and 10 days respectively. The lower one (θ_{XA} = 3 days) is selected as an example for carbon and phosphorus removal systems while θ_{XA} = 10 days -which may correspond to the sum of aerobic and anoxic SRTs when nitrification occurs- may be associated with combined (nitrogen + phosphorus) nutrient removal systems. The simulated data in Figure. 5.9 indicate that at least a one day anaerobic SRT is required in order to have EBPR in carbon and phosphorus removal systems and that phosphorus removal greatly increases with increasing anaerobic SRT. On the other hand, an anaerobic SRT at least higher than two days is required to initiate EBPR in nutrient removal systems for the selected wastewater and increasing the anaerobic SRT affects the phosphorus removal only moderately. This reflects the importance of avoiding excessive aeration and the difficulty to sustain EBPR in nutrient removal systems even when the nitrate effect is somehow eliminated.

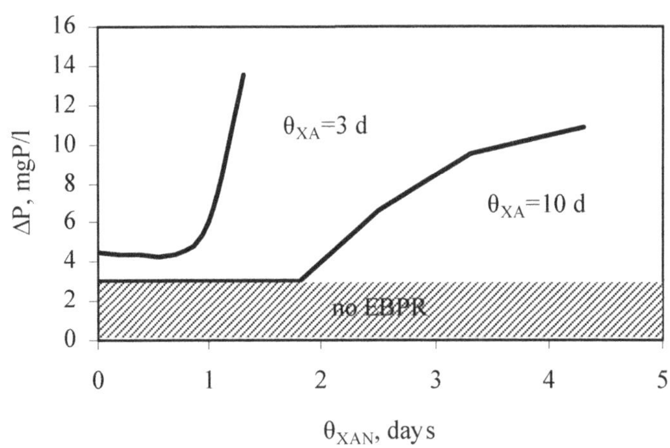

Figure 5.9. Effect of anaerobic SRT for two different aerobic SRT

5.4.2. Effect of Wastewater Composition

The COD/TP ratio of wastewater is a decisive characteristic that leads to categorizing wastewaters as carbon limited or phosphorus limited for a selected process configuration. A set of simulations was carried out to illustrate the effect of wastewater composition and namely the COD/TP ratio on system performance. A system configuration having an aerobic SRT of 10 d, anaerobic SRT of 3.3 d, total SRT of 20 d and a very low FTR (i.e. 0.2/6), was defined with the same influent COD composition and an influent TP concentration gradually decreased from 20 to 6 mg/l, corresponding to a COD/TP ratio in the range of 26.3 – 87.5, in order to assess system behaviour under carbon and phosphorus limited conditions. The results and significant characteristics of the simulation are outlined in Table 5.7. The P content of the sludge at the end of the aerobic period, as given in the Table, is calculated from the expressions below:

$$P_{cont} = \frac{i_{PBM}(X_H + X_A + X_{PAO}) + i_{PXI}X_I + X_{PP}}{X_{TSS}} \tag{5.3}$$

$$X_{TSS} = 0.9(X_H + X_A + X_{PAO}) + 0.75X_S + 0.75X_I + 0.60X_{PHA} + 3.23X_{PP} \tag{5.4}$$

where:
X_{PAO} = phosphate accumulating organisms, mg COD/l
X_{PHA} = stored poly-hydroxy-alkanoates, mg COD/l
X_{PP} = stored poly-phosphate, mg P/l
i_{PBM} = P fraction of biomass, gP/gCOD
i_{PXI} = P content of X_I, gP/gCOD

Table 5.7 Effect of influent TP and COD/TP Ratio on EBPR Performance

TP mg/l	COD/TP	An-end S_A, mg/l	S_{PO4} mg/l	ΔP mg/l	X_{PAO} / (X_H+X_{PAO}) -	X_{PP} / X_{PAO} P/COD	%P_{cont} P/TSS
20	26.3	2	8.7	11.3	0.30	0.31	6.5
15	35.5	2	3.7	11.3	0.30	0.31	6.5
12	43.8	2	1.8	10.2	0.30	0.24	5.9
10	52.5	2	1.1	8.8	0.30	0.19	5.1
8	65.6	2	0.7	7.3	0.30	0.14	4.2
6	87.5	2	0.1	5.9	0.31	0.09	3.4

The results in Table 5.7 show that maximum 11.3 mg/l of phosphorus could be removed when the influent TP is not limiting and S_A is almost completely stored as PHA in the anaerobic period. The Table also shows that the influent TP, and in turn, the poly-P content of PAOs does not affect remaining S_A concentration at the end of the anaerobic period, which indicates that the limiting process is fermentation rather than storage of S_A under these conditions. The PAO fraction of the active biomass is constant provided that COD concentration and composition are not changed as shown in Table 5.7. An increase in the influent TP results in increasing the poly-P content of PAOs until it reaches a maximum value, thereby increasing the P content of the sludge.

The effect of influent TP on phosphorus and PHA profiles within an operating cycle is illustrated in Figure 5.10. Figure 5.10.(a) shows that the influent TP affects only the PHA utilization and stored PHA is totally consumed for all influent TP values during the aerobic period due to the unnecessarily high aerobic SRT for EBPR. On the other hand, the influent TP significantly affects the S_{PO4} profile. The secondary release of phosphate is observed resulting from decay processes after depletion of PHA due to the excessive aeration as shown in Figure 5.10.(b).

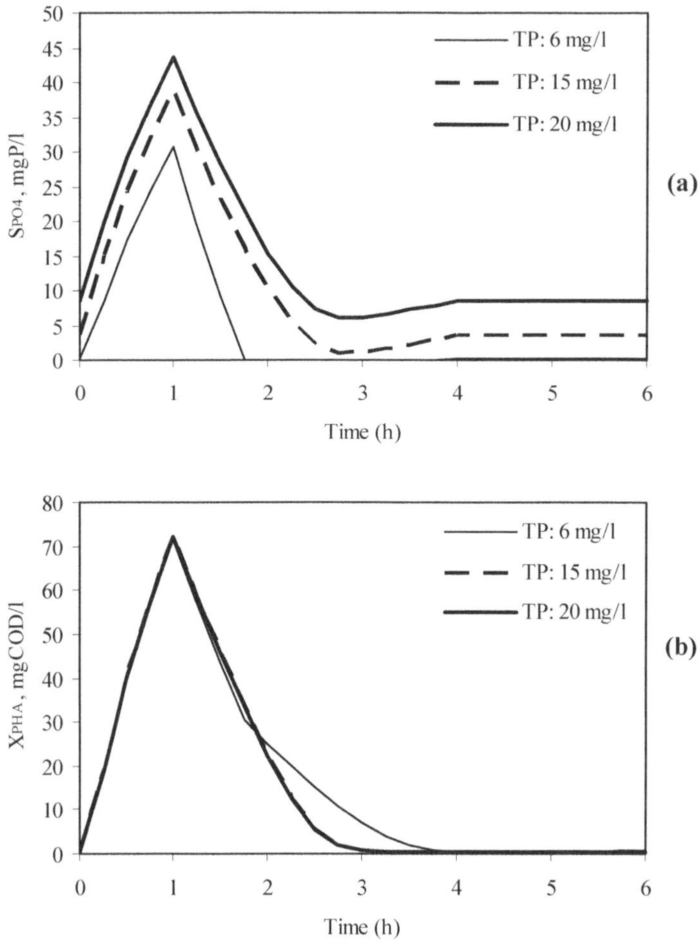

Figure 5.10. Effect of influent TP on (a) S_{PO4} profile (b) X_{PHA} profile

Interpretation of the results requires consideration of two different conditions reflecting P limitation and carbon (P storage) limitation for evaluation the effect of COD/TP ratio on EBPR. In the case of P limitation, indicating that the system still has an additional EBPR potential, increasing the TP level for a given COD concentration and composition in the influent, will not seriously affect the effluent S_{PO4} level until the existing P storage capacity is exceeded. The system will respond with a corresponding increase in the poly-P, (PP), concentration and in the P content of the activated sludge. In fact, phosphorus limited conditions cause a decrease in the P content of sludge; the effect of influent COD/TP ratio on effluent P concentration and P content of the sludge is given in Figure 5.11 which shows a decreasing trend both for S_{PO4} and P content with increasing COD/TP ratios.

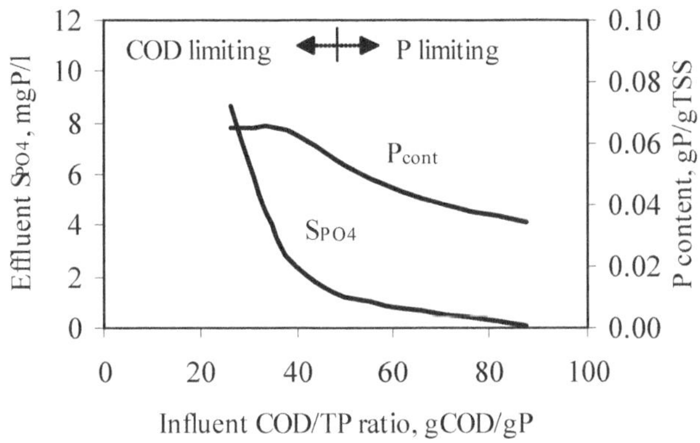

Figure 5.11. Effect of influent COD/TP ratio on ΔP

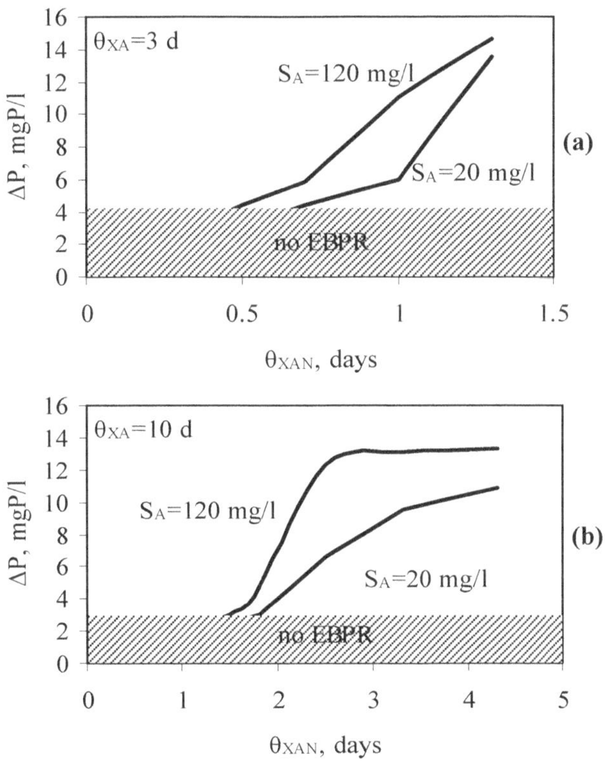

Figure 5.12. Effect of influent S_A on ΔP for (a) θ_{XA}= 3 days (b) θ_{XA}= 10 days

Since the VFA fraction of influent COD has significant effect on phosphorus removal, S_A was increased from 20 mg/l to 120 mg/l COD. Figure 5.12 (a) and (b) show that EBPR is initiated at lower anaerobic SRTs and a lower anaerobic SRT will be adequate for the same ΔP when the VFA fraction is high. This is especially important for the nutrient removal systems, which require higher aerobic (and anoxic) SRTs.

5.5. EVALUATION OF SIMULTANEOUS NITROGEN AND PHOSPHORUS REMOVAL PERFORMANCES

Nitrogen and phosphorus removal performance of the SBR process is evaluated using ASM2d similarly implemented by AQUASIM. This model is the modification of ASM2 additionally accounting for PAO growth and poly-P storage under anoxic conditions. The kinetic and stoichiometric coefficients suggested as default values at 20°C in ASM2 and the same domestic wastewater composition used for the design example in the previous chapter were adopted for the simulation runs. Simulation was carried out first with the design and operation conditions in the design example to check whether the desired effluent quality is obtained. Then, operational conditions were slightly changed to improve the design. Steady state results are outlined in the Table 5.7. Phosphorus and nitrogen profiles associated with the cyclic operation are illustrated in Figure 5.13 for the runs given in Table 5.8.

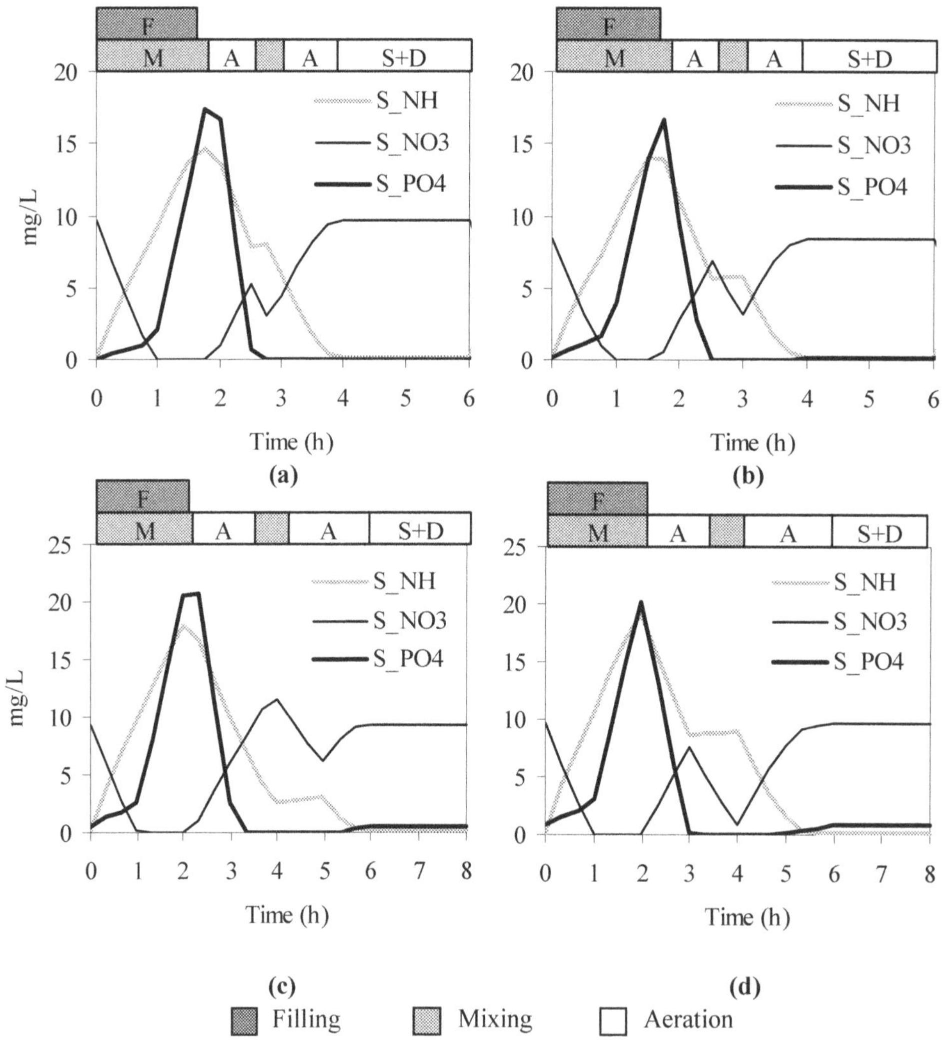

Figure 5.13. Nitrogen and phosphorus profiles. (a) Run 1 (b) Run 2 (c) Run 3 (d) Run 4

Table 5.8 Phosphorus and Nitrogen removal performances at a given operational conditions.

RUN	T_C (h)	T_{AN}/T_P	T_M/T_P	SRT (d)	V_0/V_F -	HRT (h)	θ_{XAN} (d)	T_{M2}/T_M (h/h)	Sim. S_{NO} mg/l	Sim. S_{PO4} mg/l
1	6	0.3	0.55	27.0	2.3	19.8	5.4	0.3/2.2	9.7	0.1
2	6	0.3	0.55	27.0	2.3	19.8	5.4	0.5/2.2	8.4	0.1
3	8	0.3	0.55	24.6	1.6	20.8	5.5	1.1/3.3	9.4	0.6
4	8	0.2	0.50	21.5	1.4	19.2	3.2	1.0/3.0	9.9	0.7

The first Run in Table 5.7 reflects model simulation of SBR performance operated under conditions exactly duplicating those in the design example outlined in the previous chapter, selected and calculated to meet an effluent S_{NO} of 10 mgN/l and complete phosphorus removal. As shown in the Table and plotted in Figure 5.7, the simulation results for Run 1 support for the validity of the proposed design approach as they yield an effluent S_{PO4} value of 0.1 mg/l and an S_{NO} value of 9.7 mg/l, only slightly lower than the desired effluent level.

The other simulation Runs (Runs 2-4), provide additional support for the proposed design, showing that the SBR performance remains basically the same, with only small perturbations with respect to slight modifications in selected parameters; The same design and operation conditions with only a higher second mixed period in Run 2 results in a lower effluent S_{NO} of 8.4 mg/l without an increase in the effluent S_{PO4} which implies that the operating scheme can be optimized for a more efficient N removal using simulation models, if needed. The simulation in Run 3 indicates that similar effluent quality can be obtained if T_C is selected as 8 h, but with a higher HRT of 20.8 h. On the other hand, a reduced anaerobic fraction and hence T_M/T_P ratio yielding substantially lower SRT in Run 4 is observed to lead an increase in both the effluent S_{PO4} and S_{NO}, although still in compliance with effluent limitations.

5.6. EVALUATION OF DYNAMIC CONDITIONS

Dynamic models implemented by various computer codes are valuable tools to evaluate the influence of dynamic influent and environmental conditions as well as the evaluation of system performance at cyclic steady state. Utilizing dynamic modelling to support design , operation and control of treatment plants is particularly necessary with nutrient removal processes due to their greater vulnerability to fluctuations in wastewater characteristics. In this section, the effects of temporary decrease in temperature and influent biodegradable COD concentration on effluent nutrient concentrations will be illustrated as an example.

Figure 5.14 illustrates the effect of μ_{Amax} on the S_{NH} profile of a system operating under Run D1 conditions given in Table 5.4 except μ_{Amax}. The μ_{Amax} value was decreased from 0.8 d^{-1} to 0.35 d^{-1} due to low temperature (or inhibition effect) at time zero and the initial conditions were taken from the steady state values of Run D1. The current aerobic sludge age of 8 days is not sufficient to sustain nitrification at this value of μ_{Amax} and the autotrophic biomass is bound to be washed out from the system. Figure 5.14 shows the gradual increase of the effluent S_{NH} concentration during the low temperature period of 4 days (24 cycle timc) as a result of decreasing autotrophic biomass in the reactor. It also shows that full nitrification will recover after two cycle time if μ_{Amax} resumes its original level.

The changes in COD/TKN and/or COD/TP ratios of influent may affect the effluent nutrient concentrations drastically. This effect is illustrated in Figure 5.15 (a) and (b) where the system performance was simulated over 4.5 days (18 cycle time) under the fluctuating influent biodegradable COD conditions. The system was operated with system parameters of Run 1 given in Table 5.8 and the initial values of the state variables were taken from the steady state simulation of this run fed with constant influent COD. The first two days of simulation was carried out with an influent COD pattern such that the influent biodegradable COD was 20% lower than the daily average value in two cycles per day and 20% higher in the other two cycles.

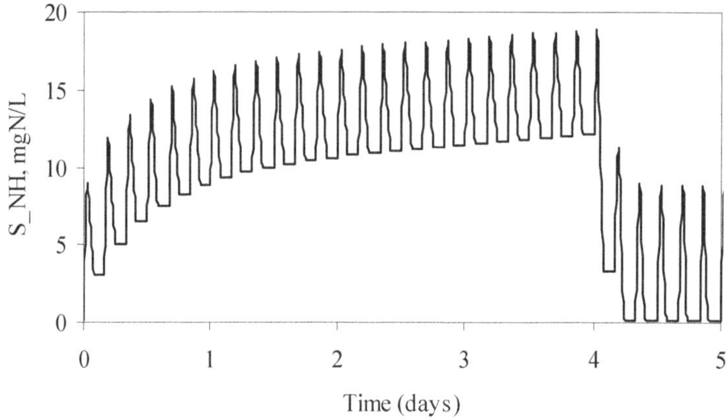

Figure 5.14. Effect of temperature on ammonia profile

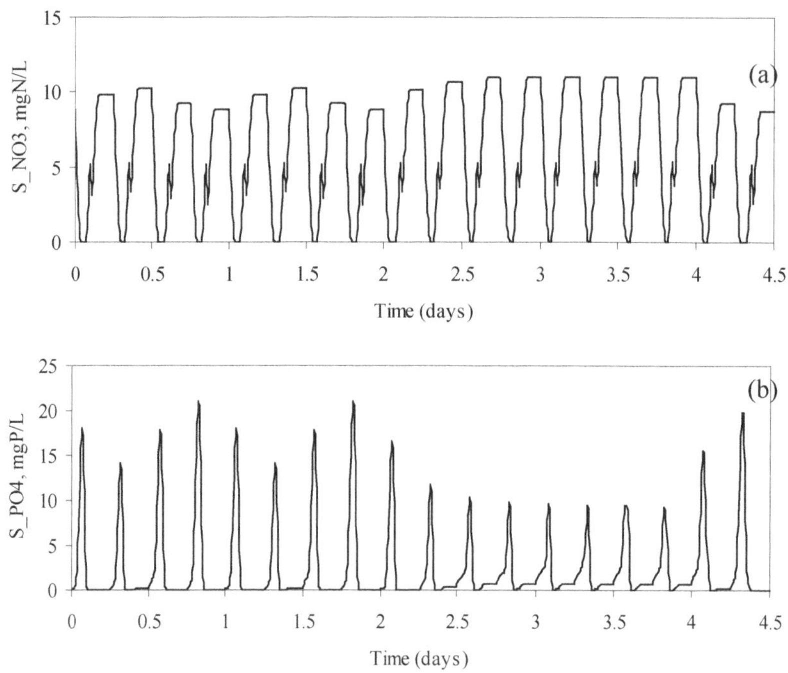

Figure 5.15. Effect of influent COD on (a) nitrate (b) phosphate profiles

Figure 5.15.(a) shows that the effluent nitrate increases up to 10.5 mgN/l under low COD conditions while it decreases to 9 mgN/l under high COD conditions. The effluent phosphate concentration increases very slightly as can be seen from Fig 5.15.(b). Then influent biodegradable COD was lowered to 315 mg/l, 30% below the average, in the following two days (8 cycle time). These conditions lead to the effluent soluble phosphate concentrations above 1 mgP/l; a frequent occurrence in nutrient removal systems in the weekend periods. Meanwhile the effluent nitrate increases up to 11 mgN/l. Simulations also show that the effluent nutrient concentrations return to the original levels in the second cycle when the influent COD resumes the previous high level.

6

Concluding Remarks

SBR is biological treatment system with many design options. Such options, when tested with a trial and error approach may prove to be quite successful in certain cases and similarly, quite detrimetal in other. Appropriate manipulation of system flexibility requires, aside from common sense, ingenuity and practical experience, full understanding and accurate interpretation of all factors affecting SBR operation for satisfactory performance.

The first issue to be visualized is the fact the suspended-growth SBR is not a novel biological treatment process but an alternative activated sludge configuration. Therefore, the great flexibility and variety of operation it offers can best be utilized on the basis of the same principles of microbial mechanisms and reactor kinetics applicable to activated sludge systems. Caution should be exercised however in setting analogies with other continuous-flow activated sludge systems: SBR is a *continuous* activated sludge system that reaches, if properly operated, a cyclic steady state and sustains an enriched biomass associated with a specific culture history for the selected wastewater characteristics and operating conditions. Therefore, the term *batch* in the name indicates nothing more than intermittent feeding and discharge. Similarly analogy of the SBR system with a continuous-flow completely mixed reactor for extended fill operation is scientifically unfounded and misleading.

Secondly, various feeding patterns do not qualify for evaluating resulting SBR operation as a set of different process offering selective treatment alternatives. In fact, it would be wrong to start the design with a pre-selected feeding pattern, if not resulting from specific physical and operation constraints. It is possible to adopt *a unified design approach* for all SBR systems operated for organic carbon , nitrogen and phosphorus removal. Such a unified design approach is also, by nature, the determining factor for the selection of the most

©2005 IWA Publishing. *Mechanism and Design of Sequencing Batch Reactors for Nutrient Removal: Scientific and Technical Report No. 19* by Nazik Artan and Derin Orhon. Published by IWA Publishing, London, UK. ISBN: 1843390825.

appropriate cyclic operation scheme, the sequence of necessary phases and filling patterns for the particular application.

One of the virtues associated with system flexibility in SBR operation is the ability of selecting the fill volume, initial volume and cycle time, all interrelated however for a given hydraulic residence time. Appropriate selection of these parameters is quite important, especially for nutrient removal systems, in view of the fact that internal recycle cannot be handled and controlled separately from sludge recycle. This major drawback in SBR operation does not allow independent control of the anoxic and anaerobic periods within the un-aerated phase.

The merit of the design mainly depends upon the level of information available for appropriate evaluation, concerning the wastewater to be treated and the biochemical mechanisms involved. This information basically involves (i) wastewater characterisation and (ii) required stoichiometric and kinetic data; it must be experimentally determined. At present, collecting the necessary information by means of scientifically acceptable and technically reliable procedures appears to be the most challenging and yet, the most overlooked aspect of SBR design.

The striking feature of a unified design approach is that it does not require elaborate model simulation, as an essential part of the procedure. It relies however on a sequence of basic expressions derived from *process stoichiometry* and *mass balance*. An integral component of the unified design basis is process modelling. An acceptably accurate model provides the necessary insight on fundamental microbial processes involved and enables translation of these processes into a design procedure.

Model simulation of SBR performance may also be quite useful in providing information on the role of selected operating parameters on system performance. Interpretation of the simulation results for process design becomes more meaningful when supported by relevant mass balance relationships between model components. While the design of carbon and nitrogen removing SBR systems may be structured upon respective mass balances with an acceptable accuracy, enhanced biological phosphorus removal, alone or coupled with nitrogen removal, necessitates a complicated array of biochemical processes involving different microbial consortia prevailing under different conditions. This makes reliable prediction system behaviour and effluent quality almost impossible without the use of simulation programs.

In this context, a basis for design is developed and presented in a stepwise approach to cover both organic carbon and nutrient removal, domestic and industrial wastewaters, strong and specific wastes, under steady-state conditions. Dynamic conditions due to temporal variations in wastewater quality and quantity, temperature fluctuations, etc., may seriously affect cyclic operation. Therefore, SBR behaviour and performance under dynamic conditions are particularly relevant for optimal system design and should be the focus of future research in this area.

Despite all claims, effect of wastewater composition and operating conditions on resulting microbial community dynamics is yet to be properly explored and understood so that it can be translated into design and process modelling. Related research, although promising, is still in its infancy. Substantial collaborative effort is needed between microbiologists and process engineers, for elucidating major questions related to population dynamics and stability for a better mechanistic understanding of SBR behaviour operated under different conditions.

References

Ardern, E. and Lockett, W.T. (1914) Experiments on the oxidation of sewage without the aid of filters. *J. Soc. Chem. Ind.* **33**, 523.

Artan, N. and Orhon, D. and Tasli, R. (2002a) Design of SBR systems for nutrient removal from wastewaters subject to seasonal fluctuations. *Wat. Sci. Technol.* **46**(8), 91-98.

Artan, N. and Tasli, R. (1999) Effect of aeration and filling patterns on nutrient removal performance in sequencing batch reactor. *Environ. Technol.* 20, 507-513.

Artan, N. Wilderer, P.A., Orhon, D., Morgenroth, E. and Ozgur, N. (2001) The mechanism and design of sequencing batch reactor systems for nutrient removal – the state of the art. *Wat. Sci. Technol.* **43**(3), 53-60.

Artan, N., Akkaya, M., Artan, S.R., (1996) Experiences with the SBR treatment of industrial wastewaters, *Proceedings, 1st IAWQ Specialized Conference on SBR Technology*, 18-20 March, Munich, Germany.

Artan, N., Orhon, D. and Choi, E. (2004) Appropriate design of activated sludge systems for nitrogen removal from high strength wastewaters. *Environ. Sci. Health A,* **39**(7), 1913-1924.

Artan, N., Ozgur Yagci, N., Artan, S. and Orhon, D. (2003) Design of SBRs for biological nitrogen removal from high strength wastewaters. *Environ. Sci. Health A,* **38**(10), 2125-2134.

Artan, N., Taşlı, R., Özgür, N and Orhon, D. (1998) The fate of phosphorus under anoxic conditions in biological nutrient removal activated sludge systems. *Biotechnol. Lett.* **20**(11), 1085-1090.

Artan, N., Wilderer, P.A., Orhon, D., Tasli, R. amd Morgenroth, E. (2002b) Model evaluation and optimization of nutrient removal potential for sequencing batch reactors. *Water SA,* **28**(4), 423-432.

Barker, P.S. and Dold, P.L. (1997) General model for biological nutrient removal activated sludge systems: model presentation. *Water Environ. Res.* **63**(5), 969-984.

Belanger, D., Bergevin, P., Lapierre, G. and Zaloum, R. (1986) Conception control and efficiency of a sequencing batch reactor for the purification of wastewater from a slaughterhouse. *Sci. Tech. Eau,* **19**, 142-156.

Bernardes, R.S. and Klapwijk, A. (1996) Biological nutrient removal in a sequencing batch reactor treating domestic wastewater. *Wat. Sci. Technol.* **33**(3), 29-38.

Brenner, A. (2000) Modeling of N and P transformations in a SBR treating municipal wastewater. *Wat. Sci. Technol.* **42**(1-2), 55-63.

Carucci, A., Chiavola, A., Majone, M. and Rolle, E. (1999) Treatment of tannery wastewater in a sequencing batch reactor, *Wat. Sci. Technol.* **40**(1), 253-259.

Dennis, R.W. and Irvine, R.L. (1979) Effect of fill:react ratio on sequencing batch biological reactors. *J. Wat. Pollut. Control Fed.* **51**(2), 255-263.

Diamandopoulos, E., Samaras, P., Dabou, X. and Sakellaropoulos, G.P. (1997) Combined treatment of landfill leachate and domestic sewage in a sequencing batch reactor. *Wat. Sci. Technol.* **36**(2-3), 61-68.

Dold, P.L., Ekama, G.A. and Marais, G.v.R. (1980) Procedures for determining influent COD fractions and the maximum growth rate of heterotrophs in activated sludge systems. *Wat. Sci. Technol.* **18**(6), 91-114.

Dollerer, J. and Wilderer, P.A. (1996) Biological treatment of leachate from hazardous waste landfills using SBBR technology. *Wat. Sci. Technol.* **34**(7-8), 437-444.

Ekama and Wentzel, (1999) Denitrification kinetics in biological N and P removal activated sludge systems treating municipal wastewaters. *Wat. Sci. Technol.* **39**(6), 69-77.

EPA (1999) Wastewater technology fact sheet–sequencing batch reactors. EPA 832-F-99-073, Office of Water, Washington D.C.

Franta, J. and Wilderer, P.A. (1997) Biological treatment of papermill wastewater by sequencing batch reactor technology to reduce residual organics. *Wat. Sci. Technol.* **35**(1), 129-136.

Goronszy, M.C. (1979) Intermittent operation of the extended aeration process for small systems. *J. Wat. Pollut. Control. Fed.* **51**(2), 274-287.

Grady C.P.L, Daigger, G.T., Lim, H.,C., (1999) *Biological Wastewater Treatment,* Marcel Dekker, Inc., New York.

Grady C.P.L., Smets, B.F., Barbeau, D.S. (1996) Variability in kinetic parameter estimates: A review of possible causes and a proposed terminology. *Water Res.* **30**(3), 742-748.

Henze, M., Grady, C.P.L., Gujer, W., Marais, G.v.R and Matsuo, T. (1987) A general model for single-sludge waste-water treatment systems. *Water Res.* **21**(5), 505-515.

Henze, M., Gujer W., Mino, T., Matsuo, T., Wentzel, M.C. and Marais, G.v.R. (1995) Activated Sludge Model. No. 2., IAWQ Scientific and Technical Report No.3. London, UK.

Henze, M., Gujer, W., Mino, T., Matsuo, T., Wentzel, M.C., Marais, G.v.R and Van Loosdrecht M.C.M. (1999) Activated Sludge Model No 2d, ASM2d. *Wat. Sci. Technol.* **39**(1), 165-183.

Hoepker, E.C. and Schroeder, E.D. (1979) The effect of loading rate on batch activated sludge effluent quality. *J. Wat. Pollut. Control Fed.* **51**(2), 264-273.

Imura, M., Suzuki, E., Kitao, T. and Iwai, S. (1993) Advanced treatment of domestic wastewater using sequencing batch reactor activated sludge process. *Wat. Sci. Technol.* **28**(10), 267-274.

Irvine and Ketchum, 1989;

Irvine, R.L., Ketchum, L.H., Breyfogle, R. and Barth, E.F. (1983) Municipal application of sequencing batch treatment. *J. Wat. Pollut. Control Fed.* **55**(5), 484-488.

Irvine, R.L., Miller, G. and Bhamrah, A.S. (1979) Sequencing batch treatment of wastewaters in rural areas. *J. Wat. Pollut. Control Fed.* **51**(2), 244-254.

Jenkins, D. and Garrison, W.E. (1968) Control of activated sludge by means of cell residence time. *J. Wat. Pollut. Control Fed.* **40**(11), Part 1.

Karahan, O., Martins, A., Orhon, D. and van Loostrecht, M.C.M. (2004) The mechanism of starch hydrolysis for substrate storage. *Water Res.* Submitted for publication.

Ketchum, L.H. and Liao, P.C. (1979) Tertiary chemical treatment for phosphorus reduction using sequencing batch reactors. *J. Wat. Pollut. Control Fed.* **51**(2), 298-304.

Kirshernman, T.L. and Hameed, S. (2000) A regulatory guide to sequencing batch reactors. *Proceedings, 2nd International Symposium on Sequencing Batch Reactor Technology*, 10-12 July, Narbonne, France.

Kuba, T., Van Loosdrecht, M.C.M., Brandse, F. And Heijen, J.J. (1997) Occurrence of denitrifyig phosphorus removing bacteria in modified UCT-type wastewater treatment plants. *Water Res.* **31**(4), 777-786.

Lee, S.I., Park, J.H., Ko, K.B. and Koopman, B. (1997) Effect of fermented swine wastes on biological nutrient removal in sequencing batch reactor. *Water. Res.* **31**(7), 1807-1812.

Ling, L. and Lo, K. V. (1999) Brewery wastewater treatment using suspended and attached growth sequencing batch reactors. *J. Environ. Sci. Health A,* **34**(2), 341-355.

Mace, S. and Mata-Alvarez, J. (2002) Utilization of SBR technology for wastewater treatment: an overview. *Ind. Eng. Chem.* **41**, 5539-5553.

McKinney, R.E. (1962) Mathematics of complete-mixing activated sludge. *Proc. Am. Soc. Civil Eng.* **88**(SA3), 87-113.

Mino, T., Liu, W.T., Kurisu, F. And Matsuo, T.(1995) Modeling glycogen storage and denitrification capability of microorganisms in enhanced biological phosphate removal processes. *Water Sci. Technol.* **31**(2), 25-34.

Mohlmann. F.W. (1917) The activated sludge method of sewage treatment. *Water survey series, No.14, University of Illinois Bulletin,* **15**, 75.

Mohseni, B.A. and Bazari, H. (2000) Biological treatment of milk factory wastewater by sequencing batch reactor. *Proceedings, 2nd International Symposium on Sequencing Batch Reactor Technology,* Volume 2, 153-156, IWA Publishing Co, London.

Murat, S., Ates Genceli, E., Tasli, R., Artan, N. and Orhon, D. (2002) Sequencing batch reactor treatment of tannery wastewater for carbon and nitrogen removal. *Wat. Sci. Technol.* **46**(9), 219-227.

Murat, S., Insel, G., Artan, N. and Orhon, D. (2004) Effect of temperature on the nitrogen removal performance of a sequencing batch reactor treating tannery wastewater. *Wat. Sci. Technol.* **48**(11), 319-326.

Orhon, D. and Artan, N. (1994) *Modelling of Activated Sludge Systems,* Technomic Publ. Co. Inc., Lancaster.

Orhon, D., Artan, N. and Cimsit, Y. (1989) The concept of soluble residual product formation in the modelling of activated sludge. *Wat. Sci. Technol.* **21**, Brighton, 339 – 350.

Orhon, D., Cimsit, Y. and Tunay, O. (1986) Substrate removal mechanism for sequencing batch reactors. *Wat. Sci. Technol.* **18**(6), 21-33.

Orhon, D., Genceli Ateş, E. and Ubay Çokgör, E. (1999) Characterization and modelling of activated sludge for tannery wastewater. *Wat. Environ. Res.* **71**(1), 50-63.

Orhon, D., Taşlı, R. and Sözen, S. (1999) Experimental basis of activated sludge treatment for industrial wastewaters – the state of the art. *Wat. Sci. Technol.* **40**(1), 1-11.

Orhon, D., Ubay Çokgör, E. and Sözen, S. (1999) Experimental basis for the hydrolysis of slowly biodegradable substrate in different wastewaters. *Wat. Sci. Technol.* **39**(1), 87-95.

Pujol, R., Lemmel, H. and Gousailles, M. (1998) A keypoint of nitrification in an uplow biofiltration reactor. *Wat. Sci. Technol.* **38**(3), 43-49.

Raper, W.G.C. and Green, J.M. (2001) Simple process for nutrient removal from food processing effluents. *Wat. Sci. Technol.* **43**(3), 123-130.

Reichert, P., Ruchti, J. and Simon, W. (1998) Aquasim 2.0, Swiss Federal Institute for Environmental Science and Technology (EAWAG), CH-8600 Dubendorf, Switzerland.

Tasli, R., Artan, N. and Orhon, D. (2001) Retrofitting SBR systems to nutrient removal in sensitive tourist areas. *Wat. Sci. Technol.* **44**(1), 121-128.

Timur, H. and Ozturk, I. (1999) Anaerobic sequencing batch reactor treatment of landfill leachate. *Water Res.* **33**(15), 3225-3230.

Torrijos, M. and Moletta, R. (1997) Winery wastewater depollution by sequencing batch reactor. *Wat. Sci. Technol.* **35**(1), 249-257.

Umble, A.K. and Ketchum, L.H.Jr. (1997) A strategy for coupling municipal wastewater treatment using the sequencing batch reactor with effluent nutrient recovery through agriculture. *Wat. Sci. Technol.* **35**(1), 177-184.

Wanner, J. (1994) *Activated sludge bulking and foaming control.* Technomic Publ. Co. Inc., Lancester.

Weber, Jr.., W.J. and DiGiano, F.A. (1995) *Process Dynamics in Environmental Systems.* 1st ed., John Wiley & Sons, Inc., New York, N.Y.

Wilderer, P.A. (1992) Sequencing batch biofilm reactor technology. In *Harnessing biotechnology for the 21st century* (ed. M.R. Ladisch and A. Bose), pp. 475-479, American Chemical Society, Washington, D.C.

Wilderer, P.A., Irvine, R.L. and Goronszy, M.C. (2001) *Sequencing Batch Reactor Technology.* Scientific and Technical Report No. 10, IWA Publishing, London UK.

Yagci, N., Artan, N., Ubay Cokgor, E., Randall, C.W. and Orhon, D. (2003) Metabolic model for acetate uptake by a mixed culture of phosphate-and glycogen-accumulating organisms under anaerobic conditions. *Biotechnol. Bioeng.* **84**(3), 359-373.

Yilmaz, G.I. and Ozturk, I. (2001) Biological ammonia removal from anaerobically pretreated landfill leachate in sequencing batch reactors (SBR). *Wat. Sci. Technol.* **43**(3), 307-314.

Zaloum, R. and Abott, M. (1997) Anaerobic pre-treatment improves single sequencing batch reactor treatment of landfill leachates. *Wat. Sci. Technol.* **35**(1), 207-214.

Index

IWA Publishing's authorised EU representative for General Product Safety Regulations is Diane D'Arras, 15 rue Duret, 75116 Paris, France, e-mail: safety@iwap.co.uk.

Printed and bound by CPI Group (UK) Ltd, Croydon, CR0 4YY
12/05/2026
02108866-0017